U.S.NRC

United States Nuclear Regulatory Commission

Protecting People and the Environment

NUREG-1808

I0488456

Sensitivity Studies of the Probabilistic Fracture Mechanics Model Used in FAVOR

Manuscript Completed: May 2006
Date Published: March 2010

Prepared by
M. EricksonKirk[1], T. Dickson[2], T. Mintz[3], F. Simonen[4]

[2]Oak Ridge National Laboratory
Oak Ridge, TN 37831-6170

[3] Southwest Research Institute
San Antonio, TX 78227-1301

[4]Pacific Northwest National Laboratory
Richland, WA 99352

[1]Office of Nuclear Regulatory Research

ABSTRACT

During plant operation, the walls of reactor pressure vessels (RPVs) are exposed to neutron radiation, resulting in localized embrittlement of the vessel steel and weld materials in the core area. If an embrittled RPV had an existing flaw of critical size and certain severe system transients were to occur, the flaw could very rapidly propagate through the vessel, resulting in a through-wall crack and challenging the integrity of the RPV. The severe transients of concern, known as pressurized thermal shock (PTS), are characterized by a rapid cooling (i.e., thermal shock) of the internal RPV surface in combination with repressurization of the RPV. Advancements in our understanding and knowledge of materials behavior, our ability to realistically model plant systems and operational characteristics, and our ability to better evaluate PTS transients to estimate loads on vessel walls led the U.S. Nuclear Regulatory Commission (NRC) to realize that the earlier analysis, conducted in the course of developing the PTS Rule in the 1980s, contained significant conservatisms.

This report, which describes sensitivity studies performed on the probabilistic fracture mechanics model, is one of a series of 21 other documents detailing the results of the NRC study.

PAPERWORK REDUCTION ACT STATEMENT

PUBLIC PROTECTION NOTIFICATION

FOREWORD

The reactor pressure vessel is exposed to neutron radiation during normal operation. Over time, the vessel steel becomes progressively more brittle in the region adjacent to the core. If a vessel had a preexisting flaw of critical size and certain severe system transients occurred, this flaw could propagate rapidly through the vessel, resulting in a through-wall crack. The severe transients of concern, known as pressurized thermal shock (PTS), are characterized by rapid cooling (i.e., thermal shock) of the internal reactor pressure vessel surface that may be combined with repressurization. The simultaneous occurrence of critical-size flaws, embrittled vessel, and a severe PTS transient is a very low probability event. The current study shows that U.S. pressurized-water reactors do not approach the levels of embrittlement to make them susceptible to PTS failure, even during extended operation well beyond the original 40-year design life.

Advancements in our understanding and knowledge of materials behavior, our ability to realistically model plant systems and operational characteristics, and our ability to better evaluate PTS transients to estimate loads on vessel walls have shown that earlier analyses, performed some 20 years ago as part of the development of the PTS rule, were overly conservative, based on the tools available at the time. Consistent with the NRC's Strategic Plan to use best-estimate analyses combined with uncertainty assessments to resolve safety-related issues, the NRC's Office of Nuclear Regulatory Research undertook a project in 1999 to develop a technical basis to support a risk-informed revision of the existing PTS Rule, set forth in Title 10, Section 50.61, of the Code of Federal Regulations [1].

Two central features of the current research approach were a focus on the use of realistic input values and models and an explicit treatment of uncertainties (using currently available uncertainty analysis tools and techniques). This approach improved significantly upon that employed in the past to establish the existing 10 CFR 50.61 embrittlement limits. The previous approach included unquantified conservatisms in many aspects of the analysis, and uncertainties were treated implicitly by incorporating them into the models.

This report is one of a series of 21 reports that provide the technical basis that the staff will consider in a potential revision of 10 CFR 50.61. The risk from PTS was determined from the integrated results of the Fifth Version of the Reactor Excursion Leak Analysis Program (RELAP5) thermal-hydraulic analyses, fracture mechanics analyses, and probabilistic risk assessment. This report documents the sensitivity studies performed on the probabilistic fracture mechanics (PFM) model used in the calculations performed to estimate PTS risk to (1) provide confidence in the robustness of the PFM model, and (2) to provide confidence that the PTS risk results for three specific plants (see NUREG-1806) can be used to develop a screening limit applicable to PWRs in general.

Brian W. Sheron, Director
Office of Nuclear Regulatory Research
U.S. Nuclear Regulatory Commission

CONTENTS

FIGURES

TABLES

EXECUTIVE SUMMARY

This report is one of a series of reports that summarize the results of a 5-year project conducted by the U.S. Nuclear Regulatory Commission's (NRC) Office of Nuclear Regulatory Research. The aim of this study was to develop a technical basis to support a revision to the pressurized thermal shock (PTS) rule found at Title 10, Section 50.61, of the *Code of Federal Regulations* [1] and the associated PTS screening criteria in a manner consistent with current NRC guidelines on risk-informed regulation. Figure ES-1, which highlights this report, also shows all of the reports that document this project.

This executive summary begins with a description of PTS, how it might occur, and what the potential consequences are for the vessel. A summary of the current regulatory approach to PTS follows, which leads directly to a discussion of the motivations for undertaking this project, and concludes with a description of how the project was conducted. This introductory material provides a context for the information presented in this report concerning the sensitivity studies the NRC performed on the probabilistic fracture mechanics (PFM) model.

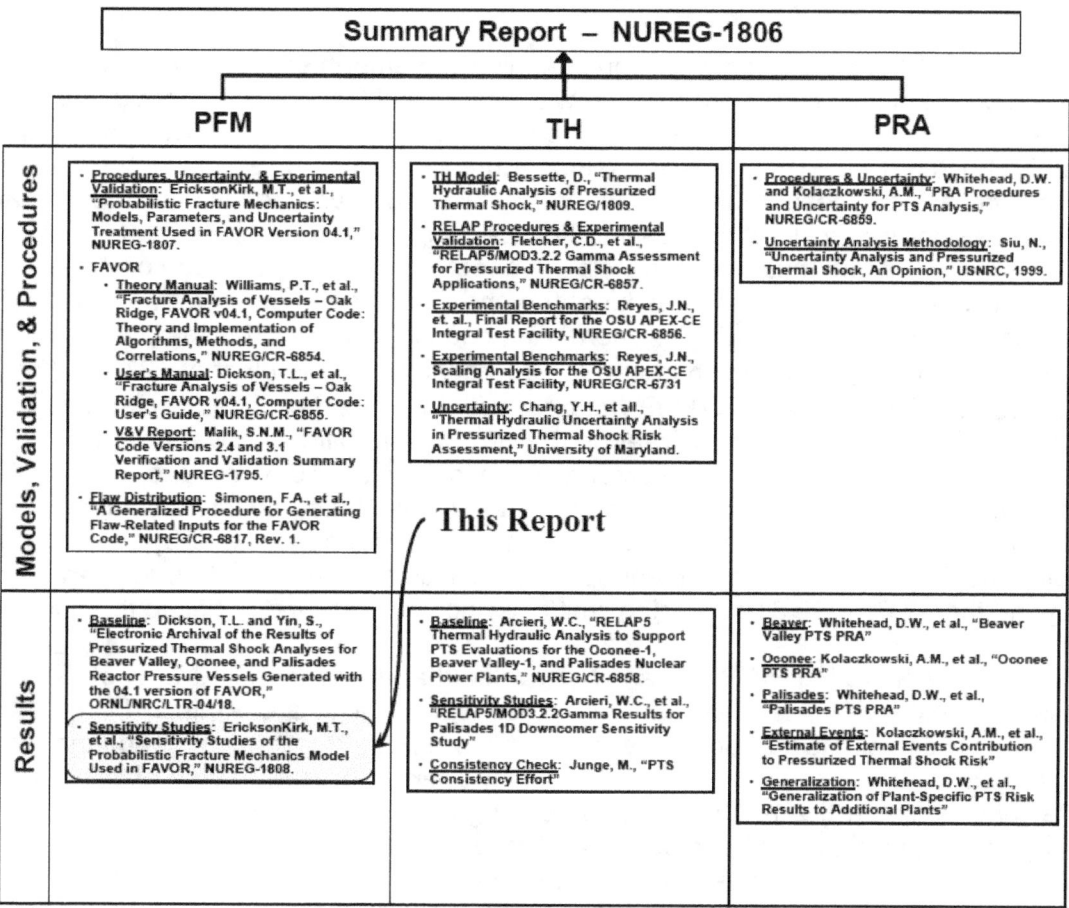

Figure ES-1 Structure of reports documenting the PTS reevaluation effort

Description of PTS

One potentially significant challenge to the structural integrity of the reactor pressure vessel (RPV) in a pressurized-water reactor (PWR) is posed by a PTS event in which rapid cooling of the downcomer occurs, possibly followed by repressurization. A number of abnormal events and postulated accidents have the potential to thermally shock the vessel (either with or without significant internal pressure); some of these include a pipe break in the primary pressure circuit, a stuck-open valve in the primary pressure circuit, and the break of the main steamline. During these events, the water level drops as a result of the contraction produced by rapid depressurization. In events involving a break in the primary pressure circuit, an additional drop in the water level occurs because of leakage from the break. Automatic systems and operators must provide makeup water in the primary system to prevent overheating of the fuel in the core. The makeup water is much colder than that held in the primary system.

The temperature drop produced by rapid depressurization, coupled with the near-ambient temperature of the makeup water, produces significant thermal stresses in the thick-section steel wall of the RPV. For embrittled RPVs, these stresses could be high enough to initiate a running crack which could propagate all the way through the vessel wall. Through-wall cracking of the RPV could precipitate core damage or, in rare cases, a large early release of radioactive material to the environment.

Current PTS Regulations

As required by 10 CFR 50.61, licensees must monitor the embrittlement of their RPVs using a surveillance program qualified in accordance with Appendix H to 10 CFR Part 50. Licensees use the results of surveillance, together with the formulae and tables in 10 CFR 50.61, to estimate the fracture toughness transition temperature (RT_{NDT}) of the steels in the vessel's beltline, as well as how these transition temperatures increase as a result of irradiation damage throughout the operational life of the vessel. For licensing purposes, 10 CFR 50.61 provides instructions on how to use these estimates of the effect of irradiation damage on RT_{NDT} to determine the value of RT_{NDT} that will occur at the end of license (EOL), a value called RT_{PTS}. Title 10, Section 50.61, of the *Code of Federal Regulations* also provides "screening limits," or maximum values of RT_{NDT}, permitted during the operating life of the plant of +270 °F (for axial welds, plates, and forgings) and +300 °F (for circumferential welds). These screening limits correspond to a limit of 5×10^{-6} events/year on the yearly probability of developing a through-wall crack [32] Should RT_{PTS} exceed these screening limits, 10 CFR 50.61 requires that the licensee either take actions to keep this value below the screening limit (i.e., by implementing "reasonably practicable" flux reductions to reduce the embrittlement rate or by de-embrittling the vessel by annealing [33]) or perform plant-specific analysis to demonstrate that operating the plant beyond the 10 CFR 50.61 screening limit does not pose an undue risk to the public [32].

While no currently operating PWR has an RT_{PTS} value that exceeds the 10 CFR 50.61 screening limit before EOL, several plants are close to the limit (3 are within 2 °F, while 10 are within 20 °F). Those plants that are close to the limit are likely to exceed it during the 20-year license renewal period that many operators currently seek. Moreover, some plants maintain their RT_{PTS} values below the 10 CFR 50.61 screening limits by implementing flux reduction (i.e., low-leakage cores, ultra-low leakage cores) and other fuel management strategies that can be economically deleterious in a deregulated marketplace. Thus, the 10 CFR 50.61 screening limits can restrict the licensable and the economic lifetime of PWRs.

Motivation for This Project

It is now widely recognized that the state of knowledge and data limitations in the early 1980s necessitated a conservative treatment of several key parameters and models used in the probabilistic calculations that provided the technical basis of the current PTS rule. The most prominent of these conservatisms include the following:

- the highly simplified treatment of plant transients (i.e., the very coarse grouping of many operational sequences (order of 10^5) into very few groups (approximately 10)) necessitated by limitations in the computational resources needed to perform multiple thermal hydraulic (TH) calculations

- the lack of any significant credit for operator action

- characterization of fracture toughness using RT_{NDT}, which has an intentional conservative bias

- the use of a flaw distribution that placed all of the flaws on the interior surface of the RPV, and, in general, contains larger flaws than those usually detected in service

- the modeling approach that treated the RPV as if it were made entirely from the most brittle of its constituent materials (welds, plates, or forgings)

- the modeling approach that estimated RPV embrittlement based only on the peak fluence occurring anywhere on the entire interior surface of the RPV

These factors indicate the high likelihood that the current 10 CFR 50.61 PTS screening limits are unnecessarily conservative. Consequently, the NRC believes that a reexamination of the technical basis for these screening limits, based on a modern understanding of all the factors that influence PTS, would most likely provide strong justification for substantial relaxation of these limits. For these reasons, the NRC Office of Nuclear Regulatory Research undertook this project with the objective of developing the technical basis to support a risk-informed revision of the PTS rule and the associated PTS screening limit.

Approach

As illustrated in Figure ES-2, there are three main models (shown as solid blue squares) that, together, permit estimation of the yearly frequency of through-wall cracking in an RPV:

(1)   a probabilistic risk assessment (PRA) event sequence analysis
(2)   a TH analysis
(3)   a PFM analysis

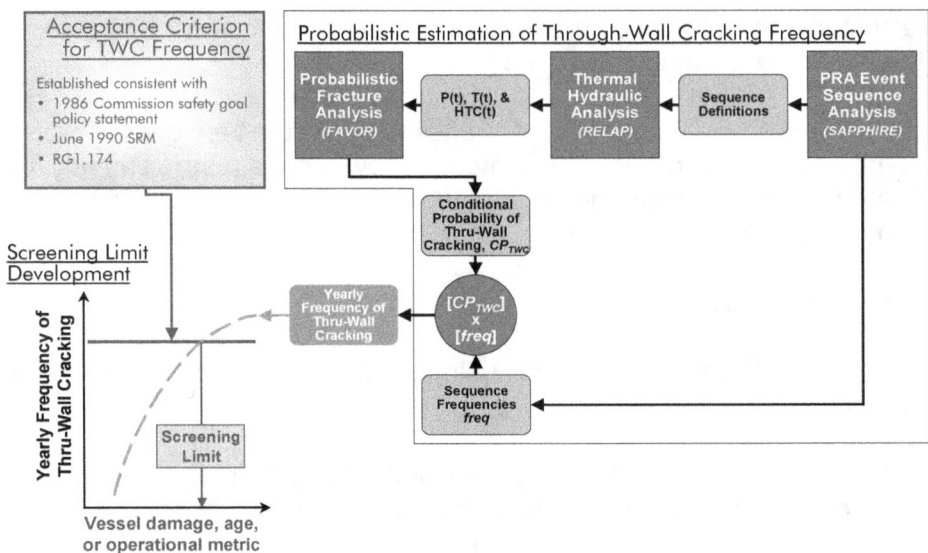

Figure ES-2 Schematic showing how a probabilistic estimate of through-wall cracking frequency (TWCF) is combined with a TWCF acceptance criterion to arrive at a proposed revision to the PTS screening limit

A PRA event sequence analysis is first performed to define the sequences of events that are likely to produce a PTS challenge to the RPV integrity, as well as to estimate the frequency with which such sequences can be expected to occur. The event sequence definitions are then passed to a TH model that estimates the temporal variation of temperature, pressure, and heat transfer coefficient in the RPV downcomer characteristic of each of the sequence definitions. These pressure, temperature, and heat transfer coefficient histories are passed to a PFM model. The PFM model uses the TH output, along with other information concerning plant design and materials of construction, to estimate the time-dependent driving force to fracture produced by a particular event sequence. The PFM model compares this estimate of fracture driving force to the fracture toughness, or fracture resistance, of the RPV steel. This comparison allows an estimation of the probability that a particular sequence of events will produce a crack all the way through the RPV wall if that sequence of events were actually to occur. The final step in the analysis involves a simple matrix multiplication of the probability of through-wall cracking (from the PFM analysis) with the frequency at which a particular event sequence is expected to occur (as defined by the event-tree analysis). This product establishes an estimate of the yearly frequency of through-wall cracking that can be expected for a particular plant after a particular period of operation when subjected to a particular sequence of events. The yearly frequency of through-wall cracking is then summed for all event sequences to estimate the total yearly frequency of through-wall cracking for the vessel due to PTS. Performance of such analyses for various operating lifetimes provides an estimate of how the yearly through-wall cracking frequency can be expected to vary over the lifetime of the plant.

The probabilistic calculations just described are performed to establish the technical basis for a revised PTS rule within an integrated systems analysis framework. The NRC approach considers a broad range of factors that influence the likelihood of vessel failure during a PTS event, while accounting for uncertainties in these factors across a breadth of technical disciplines. Two central features of this approach are (1) a focus on the use of realistic input values and models (whenever possible), and (2) an explicit treatment of uncertainties (using currently available uncertainty analysis tools and techniques). Thus, the current approach improves upon that used in the development of SECY-82-465 which included, in the many aspects of the analysis, intentional and unquantified conservatisms and which treated uncertainties implicitly by incorporating them into the models.

Key Findings

This report documents sensitivity studies the NRC performed on the FAVOR PFM model (and on PFM-related variables) with two goals in mind:

(1)   To provide confidence in the robustness of the PFM model, the NRC assessed the effect of the following credible model and input perturbations on TWCF estimates.

    —   residual stresses assumed to exist in the RPV wall
    —   embrittlement shift model
    —   resampling of chemical composition variables at the 1/4 T, 1/2 T, and 3/4 T locations for welds
    —   upper-shelf toughness model

(2)   To provide confidence that the results of the calculations for three specific plants can be generalized to apply to all PWRs, the NRC performed the following sensitivity studies to assess the influence of factors not fully considered in the baseline TWCF estimates:

    —   method for simulating increased levels of embrittlement
    —   assessment of the applicability of these results to forged vessels
    —   effect of vessel thickness

In the former category all effects were negligible or small. The small effects included the adoption of an embrittlement shift model which differs from that in American Society for Testing and Materials (ASTM) Standard E900-02 (which increases TWCF by approximately a factor of 3). The model also accounts for distinctly different copper contents in different weld layers (which reduces TWCF by approximately a factor of 2.5 relative to the assumption that the mean value of copper does not vary through the vessel thickness). Neither of these effects is significant enough to warrant a change to the baseline model or to recommend a caution regarding its robustness.

Sensitivity studies in the latter category suggest the following minor cautions regarding the general applicability of the baseline results for the three study plants reported in NUREG-1806 to all PWRs:

•   In general, the TWCF of forged PWRs can be assessed using the formulae presented in Chapter 11 of NUREG-1806 [20] by ignoring the TWCF contribution of axial welds. However, should changes in future operating conditions result in a forged vessel being subjected to very high levels of embrittlement, a plant-specific analysis to assess the effect of subclad flaws on TWCF would be warranted.

•   For PWRs with a vessel thickness ranging from 7.5 in. to 9.5 in., the TWCF values reported in NUREG-1806 [20] are realistic. These results overestimate the TWCF of the seven thinner vessels (wall thicknesses below 7 in.) and underestimate the TWCF of Palo Verde Units 1, 2, and 3, all of which have wall thicknesses above 11 in. However, these thicker vessels have very low embrittlement projected at either EOL or end of license extension, suggesting little practical effect of this underestimation.

1. INTRODUCTION

1.1 Description of Pressurized Thermal Shock

One potentially significant challenge to the structural integrity of the reactor pressure vessel (RPV) in a pressurized-water reactor (PWR) is posed by a pressurized thermal shock (PTS) event in which rapid cooling of the downcomer occurs, possibly followed by repressurization. A number of abnormal events and postulated accidents have the potential to thermally shock the vessel (either with or without significant internal pressure), including a pipe break in the primary pressure circuit, a stuck-open valve in the primary pressure circuit, and the break of the main steamline. During these events, the water level drops as a result of the contraction produced by rapid depressurization. In events involving a break in the primary pressure circuit system, an additional drop in the water level occurs because of leakage from the break. Automatic systems and operators must provide makeup water in the primary system to prevent overheating of the fuel in the core. The makeup water is much colder than that held in the primary system.

The temperature drop produced by rapid depressurization, coupled with the near-ambient temperature of the makeup water, produces significant thermal stresses in the thick-section steel wall of the RPV. For embrittled RPVs, these stresses could be high enough to initiate a running crack that could propagate all the way through the vessel wall. Through-wall cracking of the RPV could precipitate core damage or, in rare cases, a large early release of radioactive material to the environment.

1.2 PTS Limits on the Licensable Life of a Commercial Pressurized-Water Reactor

In the early 1980s, attention focused on the possibility that PTS events could challenge the integrity of the RPV wall for two reasons:

(1)   Operational experience suggested that overcooling events, while not common, did occur.

(2)   The results of in-reactor materials surveillance programs suggested that the steels used in RPV construction were prone to loss of toughness over time as the result of neutron irradiation-induced embrittlement.

The possibility of accident loading, combined with degraded material conditions, motivated investigations aimed at assessing the risk of vessel failure posed by PTS for the purpose of establishing the operational limits needed to ensure that the likelihood of RPV failures caused by PTS transients is kept sufficiently low. These efforts led to the publication of a document [38] that provided the technical basis for subsequent development of the PTS rule, found at Title 10, Section 50.61, of the *Code of Federal Regulations* [1].

As required by 10 CFR 50.61, licensees must monitor the embrittlement of their RPVs using a surveillance program qualified in accordance with Appendix H to 10 CFR Part 50. Licensees use the results of surveillance, together with the formulae and tables in 10 CFR 50.61, to estimate the fracture toughness transition temperature (RT_{NDT}[†]) of the steels in the vessel's

---

[†] The RT_{NDT} index temperature was intended to correlate with the fracture toughness transition temperature of the material. Fracture toughness and how it is reduced by neutron irradiation embrittlement are key parameters

beltline, as well as how these transition temperatures increase as a result of irradiation damage throughout the operational life of the vessel. For licensing purposes, 10 CFR 50.61 provides instructions on how to use these estimates of the effect of irradiation damage on RT_{NDT} to determine the value of RT_{NDT} that will occur at the end of license (EOL), a value called RT_{PTS}. Title 10, Section 50.61, of the *Code of Federal Regulations* also provides "screening limits," or maximum values of RT_{NDT}, permitted during the operating life of the plant of +270 °F (for axial welds, plates, and forgings) and +300 °F (for circumferential welds). These screening limits correspond to a limit of 5×10^{-6} events/year on the yearly probability of developing a through-wall crack [32]. Should RT_{PTS} exceed these screening limits, 10 CFR 50.61 requires that the licensee either take actions to keep this value below the screening limit (i.e., by implementing "reasonably practicable" flux reductions to reduce the embrittlement rate or by de-embrittling the vessel by annealing [33]) or perform a plant-specific analysis to demonstrate that operating the plant beyond the 10 CFR 50.61 screening limit does not pose an undue risk to the public [32].

While no currently operating PWR has an RT_{PTS} value that exceeds the 10 CFR 50.61 screening limit before EOL, several plants are close to the limit (3 are within 2 °F, while 10 are within 20 °F; see Figure 1-1). Those plants that are close to the limit are likely to exceed it during the 20-year license renewal period that many operators are currently seeking. Moreover, some plants maintain their RT_{PTS} values below the 10 CFR 50.61 screening limits by implementing flux reduction (e.g., low-leakage cores, ultra-low leakage cores) and other fuel management strategies that can be economically deleterious in a deregulated marketplace. Thus, the 10 CFR 50.61 screening limits can restrict the licensable and the economic lifetime of PWRs. As detailed in the next section, considerable reason exists to believe that these restrictions are not necessary to ensure public safety and, in fact, place an unnecessary burden on licensees.

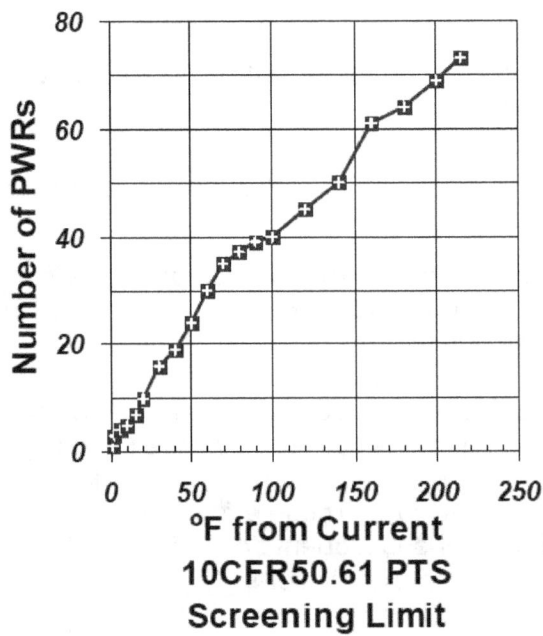

Figure 1-1 Proximity of currently operating PWRs to the 10 CFR 50.61 screening limit for PTS

---

controlling the resistance of the RPV to any loading challenge. For a more detailed description of RT_{NDT} (in specific) and fracture toughness (in general), see [19].

1.3 Technical Factors Suggesting Conservatism of the Current Rule

It is now widely recognized that state of knowledge and data limitations in the early 1980s necessitated a conservative treatment of several key parameters and models used in the probabilistic calculations that provide the technical basis [38] of the current PTS rule found at 10 CFR 50.61. The most prominent of these conservatisms include the following:

- the highly simplified treatment of plant transients (i.e., very coarse grouping of many operational sequences (order of 10^5) into very few groups (approximately 10)) necessitated by limitations in the computational resources needed to perform multiple thermal hydraulic (TH) calculations

- the lack of any significant credit for operator action

- characterization of fracture toughness using RT_{NDT}, which has an intentional conservative bias [2]

- the use of a flaw distribution that placed all of the flaws on the interior surface of the RPV, and, in general, contains larger flaws than those usually detected in service

- the modeling approach that treated the RPV as if it were made entirely from the most brittle of its constituent materials (welds, plates, or forgings)

- the modeling approach that estimated RPV embrittlement based only on the peak fluence occurring anywhere on the entire interior surface of the RPV

These factors indicate the high likelihood that the current 10 CFR 50.61 PTS screening limits are unnecessarily conservative. Consequently, the NRC believes that an examination of the technical basis for these screening limits, based on a modern understanding of all the factors that influence PTS, would most likely provide strong justification for substantial relaxation of these limits. For these reasons, the U.S. Nuclear Regulatory Commission's (NRC) Office of Nuclear Regulatory Research undertook this project with the objective of developing the technical basis to support a risk-informed revision of the PTS rule and the associated PTS screening limit.

1.4 PTS Reevaluation Project

This section describes the PTS Reevaluation Project, which the NRC Office of Nuclear Regulatory Research initiated in 1999. This section discusses restrictions placed on the model used to estimate PTS risk, the overall structure of the model, how uncertainties in the model are addressed, and how this and other reports document the results of the project.

1.4.1 Restrictions on Model

This research effort seeks to establish the technical basis for a new PTS screening limit. To enable all operators of commercial PWRs to assess the state of their RPV relative to such a new criterion, without making new material property measurements, the fracture toughness properties of the RPV steels need to be estimated using only information that is currently available (i.e., RT_{NDT} values, upper-shelf energy values, and the chemical composition of the beltline materials). The NRC Reactor Vessel Integrity Database [31] summarizes all of this information.

1.4.2 Overall Structure of Model

The overall model involves the following three major components, which are illustrated (along with their interactions) in Figure 1-2:

Component 1—Probabilistic Evaluation of Through-Wall Cracking Frequency. Estimate frequency of through-wall cracking as a result of a PTS event given the operating, design, and material conditions in a particular plant.

Component 2—Acceptance Criterion for Through-Wall Cracking Frequency. Establish a value of reactor vessel failure frequency (RVFF) consistent with current guidance on risk-informed decisionmaking.

Component 3—Screening Limit Development. Compare the results of the two preceding steps to determine if some simple, materials-based screening limit for PTS can be established. Conceptually, plants falling below the screening limit would be deemed adequately resistant to a PTS challenge and would not require further analysis. Conversely, more detailed, plant-specific analysis would be needed to demonstrate the safety of plant operation at embrittlement levels beyond the screening limit.

Each of these components is described in the following subsections.

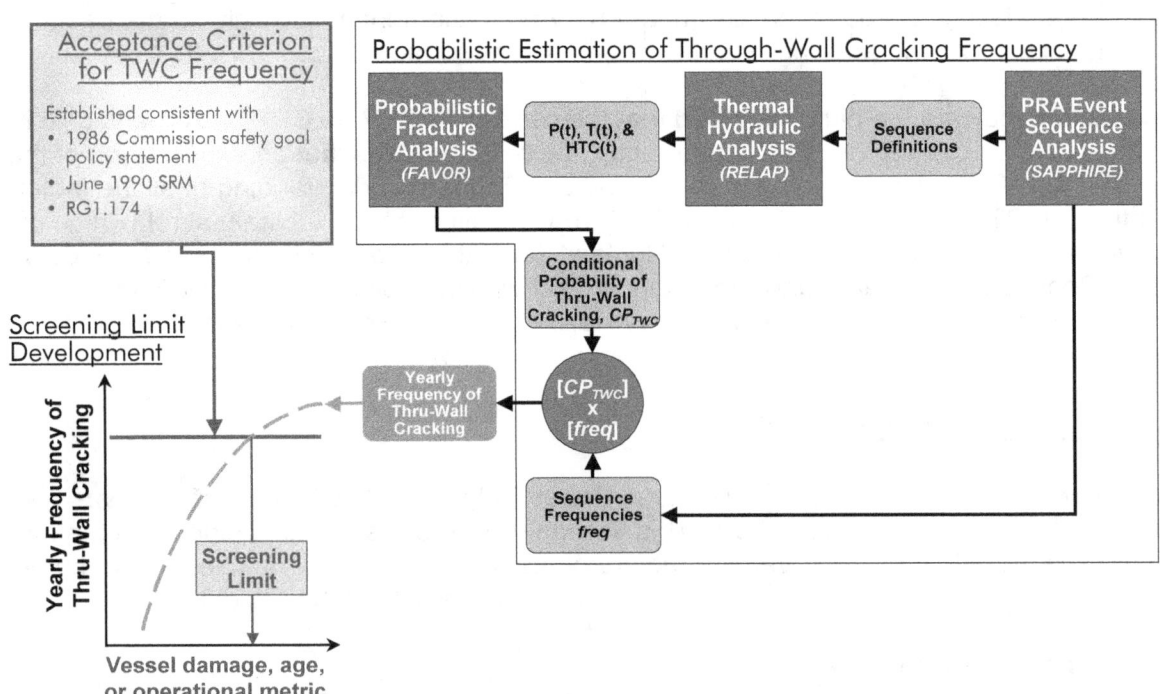

Figure 1-2 High-level schematic showing how a probabilistic estimate of through-wall cracking frequency (TWCF) is combined with a TWCF acceptance criterion to arrive at a proposed revision to the PTS screening limit

1.4.2.1 Component 1—Probabilistic Estimation of Through-Wall Cracking Frequency

As illustrated in Figure 1-2, there are three main models (shown as solid blue squares) that, together, permit estimation of the yearly frequency of through-wall cracking in an RPV:

(1)   a probabilistic risk assessment (PRA) event sequence analysis
(2)   a TH analysis
(3)   a PFM analysis

The following subsections first describe these three models in general, and then describe their sequential execution to provide the reader with an appreciation for the interrelationships and interfaces between the different models (Section 1.4.2.1.1). Next, the subsections describe the iterative process the NRC undertook, which involved repeated execution of all three models in sequence, to arrive at final models for each plant (Section 1.4.2.1.2). Finally, the subsections discuss the three specific plants the NRC analyzed in detail (Section 1.4.2.1.3). This section concludes with a discussion of the steps taken to ensure that the conclusions based on these three analyses apply to domestic PWRs in general (Section 1.4.2.1.4).

1.4.2.1.1 Sequential Description of How PRA, TH, and PFM Models Are Used to Estimate TWCF

A PRA event sequence analysis is first performed to define the sequences of events that are likely to produce a PTS challenge to RPV integrity, as well as to estimate the frequency with which such sequences can be expected to occur. The event sequence definitions are then passed to a TH model that estimates the temporal variation of temperature, pressure, and heat transfer coefficient in the RPV downcomer characteristic of each of the sequence definitions. These pressure, temperature, and heat transfer coefficient histories are passed to a PFM model. The PFM model uses the TH output, along with other information concerning plant design and materials of construction, to estimate the time-dependent driving force to fracture produced by a particular event sequence. The PFM model compares this estimate of fracture driving force to the fracture toughness, or fracture resistance, of the RPV steel. This comparison allows an estimation of the probability that a particular sequence of events will produce a crack all the way through the RPV wall if that sequence of events were actually to occur. The final step in the analysis involves a simple matrix multiplication of the probability of through-wall cracking (from the PFM analysis) with the frequency at which a particular event sequence is expected to occur (as defined by the event-tree analysis). This product establishes an estimate of the yearly frequency of through-wall cracking that can be expected for a particular plant after a particular period of operation when subjected to a particular sequence of events. The yearly frequency of through-wall cracking is then summed for all event sequences to estimate the total yearly frequency of through-wall cracking for the vessel due to PTS. Performance of such analyses for various operating lifetimes provides an estimate of how the yearly through-wall cracking frequency can be expected to vary over the lifetime of the plant.

1.4.2.1.2 Iterative Process Used to Establish Plant-Specific Models

A PRA event-tree approach was used to identify the set of transients which represent a particular plant, wherein many thousands of different initiating event sequences are "binned" together into groups of transients believed to produce similar TH outcomes (i.e., similar variations of temperature, pressure, and heat transfer coefficient versus time in the downcomer). Characteristics, such as similarity of break size, similarity of operator action, etc., guide judgments regarding what transients to put into what bin, resulting in "bins" such as medium break primary system loss-of-coolant accidents (LOCAs), main steamline breaks, etc. From each of the tens or hundreds of individual event sequences in each bin, a single sequence was then selected and programmed into the TH code RELAP to define the variation of pressure, temperature, and heat transfer coefficient versus time. These TH transient definitions were then passed to the PFM code FAVOR, which estimated the conditional probability of through-wall cracking (CPTWC) for each transient. When multiplied by the initiating event frequency estimates from PRA, these CPTWC become through-wall cracking frequency (TWCF) values, which, when rank ordered, estimate the degree to which each bin contributes to the total TWCF of the vessel. At this stage many bins are found to contribute very little or nothing at all to the TWCF, and so receive little further scrutiny. However, some bins invariably dominate the TWCF estimate. These bins are then further subdivided by partitioning the initiating event frequency of the bin, and by selecting a TH transient to represent each part of the original bin. FAVOR then reanalyzes this refined model and the bins that provide significant contributions to TWCF are again examined. This process of bin partitioning and selection of a TH transient to represent each newly partitioned bin continues until such time as the total estimated TWCF for the plant no longer changes significantly.

1.4.2.1.3 Plant-Specific Analyses Performed

In this project, the NRC performed detailed calculations for three operating PWRs (Beaver Valley 1, Oconee 1, and Palisades; see Figure 1-3. Together, the three plants represent a wide range of design and construction methods, and they contain some of the most embrittled RPVs in the operating fleet.

1.4.2.1.4 Generalization to All Domestic PWRs

Because the objective of this project is to develop a revision to the PTS screening limit expressed in 10 CFR 50.61 that applies to PWRs in general, the extent to which these three plant-specific analyses adequately address (in either a representative or bounding sense) the range of conditions experienced by all domestic PWRs must be understood. To achieve this goal, the NRC has taken the following actions:

- High embrittlement plant
- Westinghouse design

- High embrittlement plant
- Combustion Engineering design

- Plant used in 1980s PTS study
- Babcox & Wilcox design

Figure 1-3 The three plants analyzed in detail in the PTS reevaluation effort

- The NRC performed sensitivity studies on both the TH and PFM models to address the effect of credible changes to the model and/or its input parameters. The results of these studies provide insights regarding how robust the conclusions based on three plants are when applied to the entire PWR population.

- The NRC examined the plant design and operational characteristics of five additional plants. The aim of this analysis was to determine whether the design and operational features identified as being important in the three plant-specific analyses vary significantly enough in the general population to question the generality of the results.

- In the three plant-specific analyses, the NRC assumed that the only possible causes of PTS events have origins that are internal to the plant. However, external events, such as fires, floods, earthquakes, can also be PTS precursors. The NRC therefore examined the potential for external initiating events to create significant additional risk relative to the internal initiating events which have already been modeled in detail.

1.4.2.2 Component 2—Acceptance Criterion for Through-Wall Cracking Frequency

Since the issuance of SECY-82-465 and the publication of the original PTS rule, the NRC has established a considerable amount of guidance on the use of risk metrics and risk information in regulation (e.g., the Safety Goal Policy Statement, the PRA Policy Statement, and RG 1.174). To ensure consistency of the PTS reevaluation project with this guidance, the staff identified and assessed options for a risk-informed criterion for the RVFF (currently specified in RG 1.154 in terms of TWCF).

As described in 39, the options developed involved qualitative concerns (e.g., the definition of RPV failure) and quantitative concerns (e.g., a numerical criterion for the RVFF). The options reflected uncertainties in the margin between PTS-induced RPV failure, core damage, and large early release. The options also incorporated input received from the Advisory Committee on Reactor Safeguards (ACRS) [27] regarding concerns about the potential for large-scale oxidation of reactor fuel in an air environment.

The assessment of the options involved identification of technical issues unique to PTS accident scenario development, development of an accident progression event tree to structure consideration of the issues, performance of a scoping study of the issue of containment performance during PTS accidents, and review of the options in light of this information. The scoping study involved collection and evaluation of available information, performance of a few limited-scope TH and structural calculations, and a semi-quantitative analysis of the likelihood of various accident progression scenarios.

1.4.2.3 Component 3— Screening Limit Development

As illustrated schematically in Figure 1-2 (lower left corner), a screening limit for PTS can be established based on a simple comparison of estimates of the TWCF as a function of an appropriate measure of RPV embrittlement with the RVFF acceptance criterion (or RVFF*). Beyond the work needed to establish both the TWCF versus embrittlement curve and RVFF* values, it is also necessary to establish a suitable vessel damage metric that, ideally, allows different conditions in different materials at different plants to be normalized. From a practical standpoint, "suitable" implies that the metric needs to be based only on information regarding plant operation and materials that is readily available.

1.4.3 Uncertainty Treatment

At the outset of this project in 1999, the NRC staff reviewed the existing approach for PRA modeling, focusing on how uncertainties should be treated, how they were propagated through the PRA, TH, and PFM models, and how the approach compared with the NRC guidelines on work supporting risk-informed regulation [41]. This review established a general framework for model development and uncertainty treatment, which is summarized in the following paragraphs.

This project included the performance of probabilistic calculations to establish the technical basis for a revised PTS rule within an integrated systems analysis framework [57]. The NRC approach considers a broad range of factors that influence the likelihood of vessel failure during a PTS event, while accounting for uncertainties in these factors across a breadth of technical disciplines [41]. Two central features of this approach are (1) a focus on the use of realistic input values and models (whenever possible), and (2) an explicit treatment of uncertainties (using currently available uncertainty analysis tools and techniques). Thus, the current approach improves upon that used in the development of SECY-82-465 which included, in the many aspects of the analysis, intentional and unquantified conservatisms and which treated uncertainties implicitly by incorporating them into the models (e.g., RT_{NDT}).

The probabilistic models distinguish between two types of uncertainties—aleatory and epistemic. Aleatory uncertainties arise from the randomness inherent to a physical or human process, whereas epistemic uncertainties are caused by a limitation in the current state of knowledge (or understanding) of that process. A practical way to distinguish between aleatory and epistemic uncertainties is that epistemic uncertainties can, in principle, be reduced by an increased state of knowledge. Conversely, because aleatory uncertainties arise from the randomness at a level below which a particular process is modeled, they are fundamentally irreducible. The distinction between aleatory and epistemic uncertainties is an important part of the PTS analysis because different mathematical and/or modeling procedures are used to represent these different uncertainty types.

1.4.4 Project Documentation

This report is one of a series of reports that summarize the results of a PTS reevaluation project. Figure 1-4 illustrates the overall structure of this documentation and highlights this report.

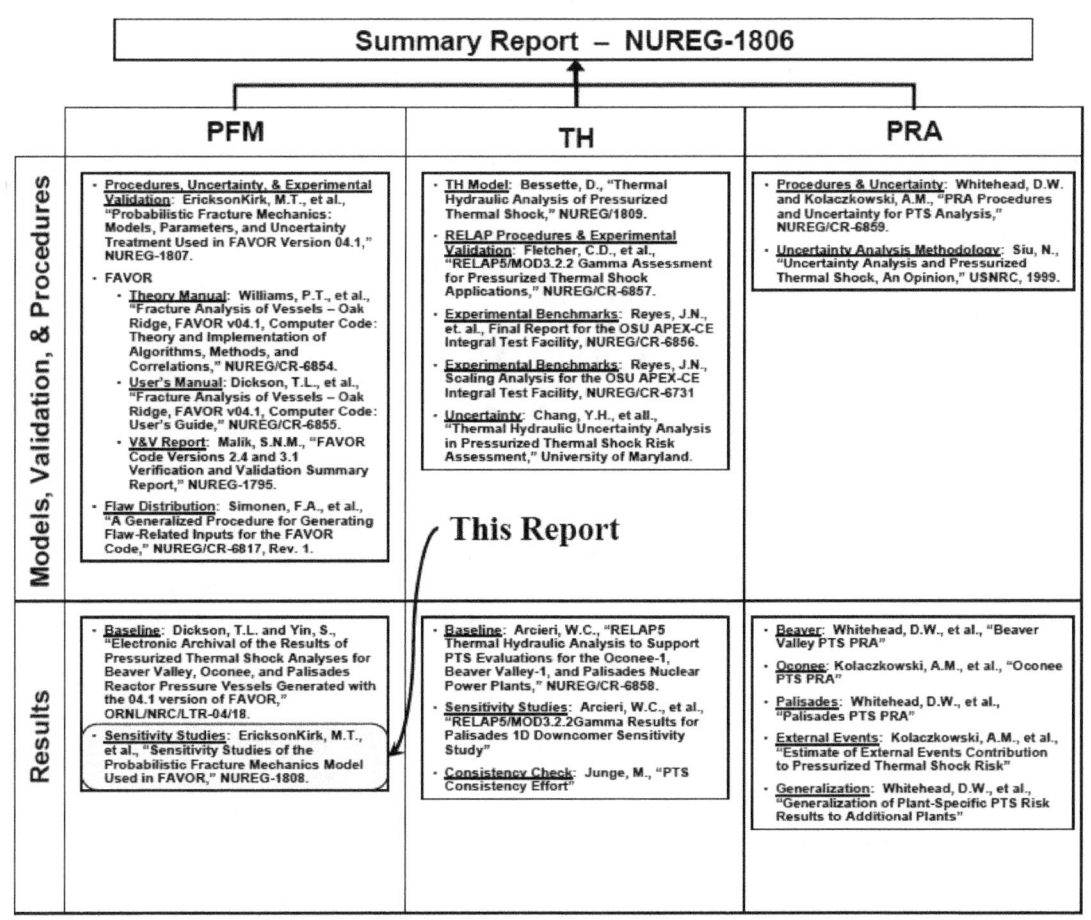

Figure 1-4 Structure of reports documenting the PTS reevaluation effort

2. OBJECTIVE, SCOPE, AND STRUCTURE OF THIS REPORT

Other reports describe the details of the PFM model and its implementation into the FAVOR computer code [19, 56]. Additionally, the baseline TWCF results for three of the study plants (Oconee, Beaver Valley, and Palisades) have been reported (see Figure 2-1 and References 20, 15). This report documents sensitivity studies performed on the FAVOR PFM model (and on PFM-related variables) with two goals in mind:

(1)   To provide confidence in the robustness of the PFM model, the NRC assessed the effect of credible model and input perturbations on TWCF estimates.

(2)   To provide confidence that the results of the calculations for three specific plants can be generalized to apply to all PWRs, the NRC performed sensitivity studies to assess the influence of factors not fully considered in the baseline TWCF.

The remainder of this document is structured as follows:

*   Chapter 3 describes a systematic process for identifying which parts of the PFM model should be subjected to a sensitivity study (and which parts should not) to provide confidence in the robustness of the PFM model.

*   Chapter 4 applies the process developed in Chapter 3 to define the sensitivity studies that will be performed.

*   Chapter 5 discusses the sensitivity studies needed to provide confidence that the results of the calculations for the three specific study plants apply in general to all PWRs.

*   Chapter 6 presents the study conclusions.

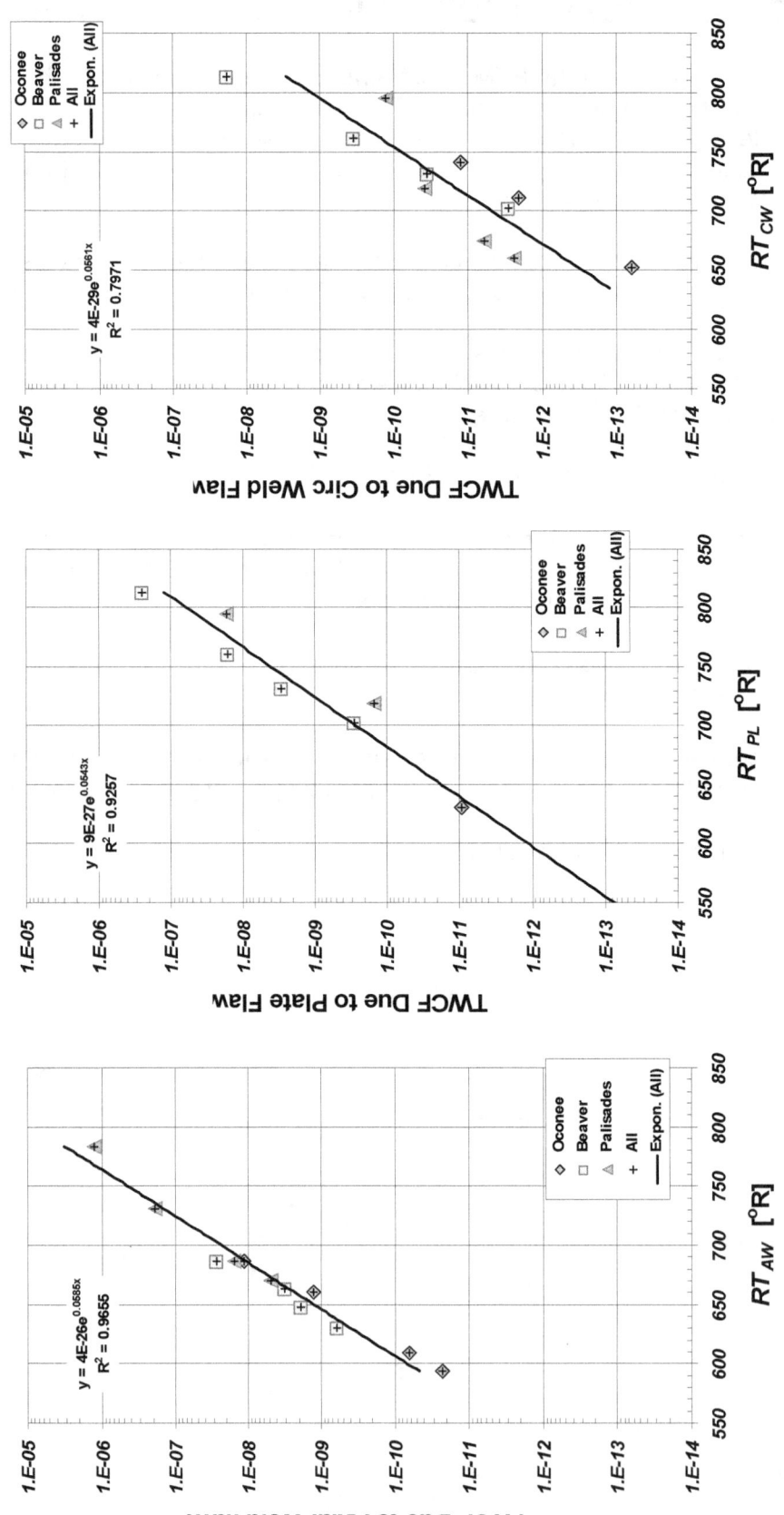

Figure 2-1 Correlation of through-wall cracking frequencies with reference temperature
metrics for the three study plants (°R = °F + 459.69) reported in NUREG-1806 [20]

2-2

3. PROCESS FOR IDENTIFYING SENSITIVITY STUDIES TO ASSESS THE ROBUSTNESS OF THE PFM MODEL

3.1 Ideas Underlying the Process

A detailed depiction (Figure 3-1) of the model used to generate predictions of through-wall cracking frequency (TWCF) for the three study plants reveals it to be a complex combination of a large number of submodels and parameter inputs. Figure 3-1 illustrates how these submodels (blue symbols) and parameter inputs (beige, round-corner rectangles) combine to produce intermediate calculated results that, upon passing through yet more submodels, eventually become an estimated distribution of TWCF. The existence of each submodel and parameter input in the PFM model, and their arrangement with respect to one another, represents a decision to structure the overall model in a particular way. Changing any one of these decisions can, in principle, change the output of the model (i.e., the distribution of TWCF values). Therefore, in this report the NRC investigates the degree to which the selection of credible alternative submodels may influence the TWCF estimates. Additionally, many of the input parameters to the PFM cannot be known precisely. Therefore, the NRC also investigates the degree to which credible variations in the input parameters change the TWCF estimates.

This approach of basing sensitivity studies on credible alternative submodels and/or on credible variations of the input parameters follows directly from two principles of the overall approach to model building (see Section 1.4.3):

- the use of realistic input values and submodels
- an explicit treatment of uncertainties

These principles have permitted calculation of TWCF estimates that are systematically biased neither high nor low (i.e., values that represent a "best estimate" to the greatest extent practicable). By basing sensitivity studies on credible alternative submodels and on credible variations of the input parameters, the NRC maintains these principles, thereby allowing the TWCF estimates to maintain their "best estimate" label. It can be noted that the model explicitly accounts for the sensitivity of the many input variables to credible variations of their values by sampling from probability distributions that have been established based on data [19, 56]. Finding credible alternative data upon which to justify a sensitivity study can be difficult in this situation.

The focus on motivating sensitivity studies based only on credible perturbations to the baseline model deviates from that taken previously [38], wherein sensitivity studies were either focused on "important" parameters and submodels (i.e., those to which the TWCF was believed to be sensitive), or were performed seemingly without consideration of either the technical justification for the baseline submodel nor the credibility of the alternative submodel used to motivate the sensitivity study. In most cases, it is important to avoid such *ad hoc* justifications for performing sensitivity studies. Low sensitivity of the output TWCF to a change in a submodel or input having an inadequate technical justification does not provide a rational basis for accepting that submodel or input as part of the overall model. Similarly, high sensitivity of the output to a well-justified submodel or input provides neither a basis for condemning that submodel/input nor for adopting arbitrary margins in an effort to compensate for the high sensitivity.

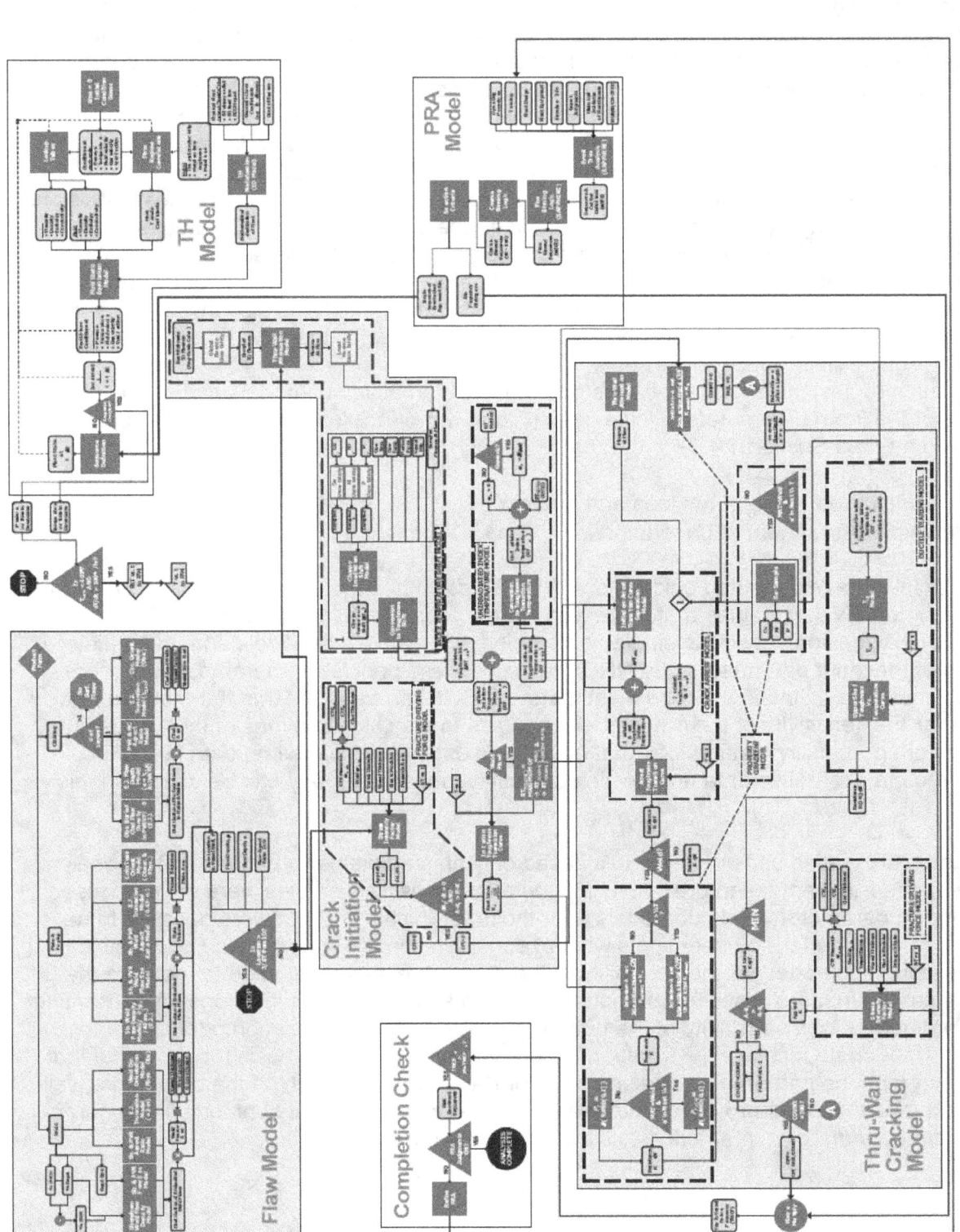

Figure 3-1 Schematic illustration of overall model used in the PTS reevaluation project

3-2

Table 3-1 Model and Parameter Classification Scheme

Color	For Models	For Parameters
M1	A correct and credible representation of the underlying physical process	The value (for deterministic parameters) or the mean and distribution (for stochastic parameters) accurately represent the parameter for the conditions of interest.
M1	The applicability of this model in its application to all conditions of interest cannot be reliably assessed with the current state of knowledge.	N/A
M1	Alternatives to the model adopted exist. These alternatives have roughly equal justification to the model adopted.	N/A
M1	A model providing a conservative representation of the underlying physical process	The value (for deterministic parameters) or the mean and distribution (for stochastic parameters) conservatively represent the parameter for the conditions of interest.
M1	While it is believed that the model provides a correct and credible representation of the underlying physical process, when uncertainties arose in constructing the model conservative decisions were made.	N/A

3.2 Process for Sensitivity Study Identification

Table 3-1 provides a set of classifications used to assess the correctness and physical credibility of the various submodels, and the accuracy of the various parameter inputs in the PFM model depicted in Figure 3-1. These classifications allow one to make systematic and consistent decisions in the selection of what sensitivity studies should be performed, as described below:

M1   **NO**—There is little benefit in subjecting a correct submodel or an accurate parameter to a sensitivity study. Instead of performing a sensitivity study, documentation is provided describing why a sensitivity study is not warranted.

M1   **USUALLY**—Because of limitations on available information, the correctness of these submodels is unknown. This state of knowledge limitation also restricts the ability to develop credible alternative models that could motivate a sensitivity study. Nevertheless, it may be important to conduct sensitivity studies on these submodels to ensure that the engineering decisions made in developing the submodel do not have unforeseen significant effects on the estimated TWCF values.

M1 **USUALLY**—Because alternative submodels to that used in the overall model exist, and these alternatives models have roughly equal justification, these submodels are usually candidates for sensitivity study.

M1 **MAYBE**—Conservative submodels and parameter inputs have been adopted as part of the overall model either because no more accurate information is available to support a better decision or because adopting a conservative submodel or parameter input somehow enhances the applicability of the results to all PWRs. In this situation, sensitivity studies are generally not useful, however, the merits of each case are discussed and documented individually.

M1 **MAYBE**—If a quantifiable and credible basis can be developed to support a less conservative model, or if independent information becomes available suggesting an alternative model, then this information provides the basis of a sensitivity study.

As suggested by these descriptions, an advanced state of knowledge may motivate very specific sensitivity studies to examine the effect of, for example, adopting a particular alternative model. In other situations in which the state of knowledge is not so advanced, specifically for YELLOW and RED models and parameters, more arbitrary sensitivity studies may be conducted simply to examine what effect the changing of an input value or model has on the estimated TWCF.

The results of these sensitivity studies are described according to the terms illustrated in Figure 3-2:

- <u>Mitigated</u>—Output changes are smaller than input changes.
- <u>Proportional</u>—Output changes are of the same order as input changes.
- <u>Magnified</u>—Output changes are larger than input changes.

The next chapter uses this classification scheme to identify what sensitivity studies will and will not be performed.

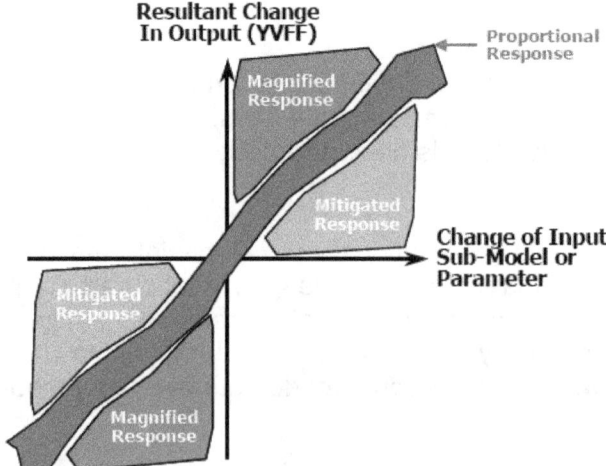

Figure 3-2 Illustration of potential outcomes of sensitivity studies

4. DEFINITION OF SENSITIVITY STUDIES PERFORMED TO ASSESS THE ROBUSTNESS OF THE PFM MODEL

Figure 4-1 depicts the overall model the NRC used in the PTS reevaluation effort with submodels and parameters classified according to the criteria outlined in Table 3-1. While both the PRA model and the TH model appear in Figure 4-1, other reports in this series [55 and 6, respectively] detail the sensitivity studies performed on these models. This report addresses the sensitivity studies performed on the PFM portion of the model. These are discussed in the following sections, subdivided by the three major parts of the PFM model:

- Section 4.1 describes the need for sensitivity studies on the flaw model.

- Section 4.2 describes the need for sensitivity studies on the crack initiation model.

- Section 4.3 describes the need for sensitivity studies on the through-wall cracking model.

4.1 Flaw Model

The flaw model provides estimates of the density (flaws per unit area or volume), size, and location in the vessel wall of initial fabrication defects[‡]. This flaw distribution, reported in detail by Simonen, represents a major improvement in realism relative to that adopted in previous studies of PTS risk [40]. Indeed, one of the major unknowns/uncertainties identified in the last comprehensive evaluation of PTS [38] was the distribution of flaws assumed to exist in the RPV wall. SECY-82-465 used flaw models based on the Marshall study, which included data from a limited population of nuclear vessels and from many non-nuclear vessels [25]. These flaw measurements were part of routine pre-service nondestructive examinations (NDE) performed 25 or more years ago at vessel fabrication shops. Because of the limitations of the NDE technology available at the time, the Marshall flaw distribution only provides a reasonable representation for flaws having depth dimensions larger than approximately 1 in. Nevertheless, SECY-82-465 and the IPTS studies [28, 29, 30] applied the Marshall distribution by extrapolating fits to the data to the much smaller flaws of concern in PTS calculations (less than approximately 1/4 in.). Additionally, all flaws in the Marshall distribution were assumed to break the inner diameter surface of the RPV despite the fact that the observations rarely, if ever, revealed surface-breaking flaws in nuclear-grade construction.

Table 4-1 Summary of Sources of Experimental Data Sources for the Flaw Distribution

Vessel	Weld	Plate	Clad
PVRUF	9150	855	1650
Shoreham	10375	975	--
Hope Creek	245	550	--
River Bend	2440	1465	--
Table entries represent volume of material examined in in³.			

[‡] Growth of initial fabrication defects resulting from subcritical cracking mechanisms does not need to be considered; see Chapter 3 of Reference 19.

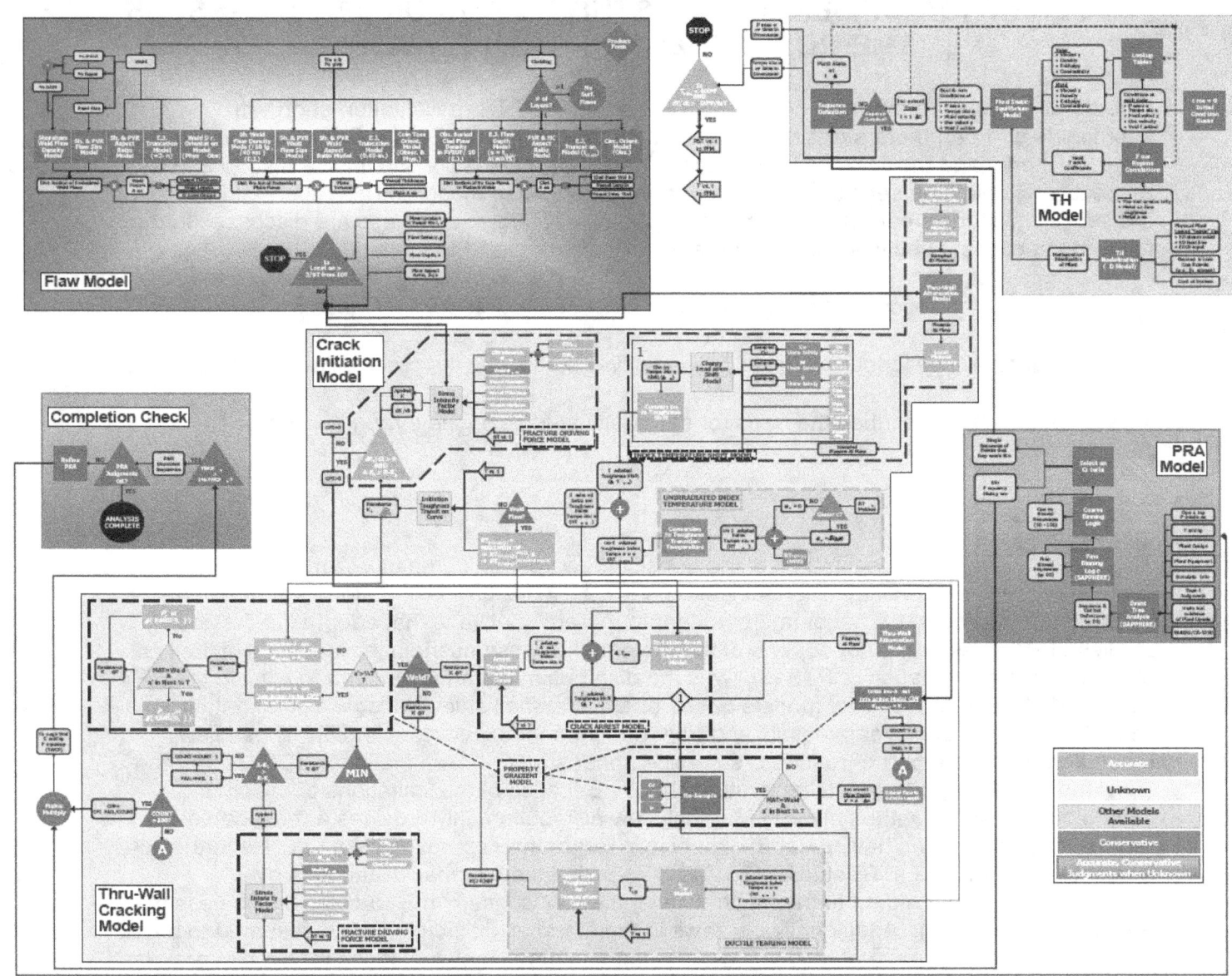

Figure 4-1 Classification of the submodels and parameter inputs used in the PTS reevaluation project according to Table 3-1 criteria

Table 4-1 summarizes the various sources of experimental data Simonen et al. used to develop the flaw distributions employed in FAVOR. While the volume of material represented in Table 4-1 improves greatly on the Marshall flaw distributions [25], an inescapable conclusion is also that the quantity of available data is also quite small compared with the volume of RPV material in service. Consequently, it is not possible to ensure on an empirical basis alone that the flaw distributions developed based on these data apply to all PWRs in general. However, the flaw distributions proposed in Simonen rely on the experimental evidence gained from inspections of the materials summarized in Table 4-1 and do not rest solely on this empirical evidence [40]. Along with empirical data, Simonen et al. used both physical models and expert opinions when developing their recommended flaw distributions. Additionally, where detailed information was lacking, Simonen et al. made systematically conservative judgments including the following:

- All NDE indications used in constructing the flaw models were treated as cracks and, therefore, potentially deleterious to RPV integrity. However, many of these indications were in fact volumetric, which lessens significantly the probability of brittle fracture initiation.

- For the larger weld flaws that dominate the estimated failure probabilities, the density of flaws found in the Shoreham vessel was adopted because it exceeds the density observed in the Pressure Vessel Research Users Facility (PVRUF) weld, the only other weld that was extensively characterized.

- FAVOR imposes truncation limits on the sizes of flaws in plates, and the flaws in welds exceed in size the largest flaw found in either PVRUF or Shoreham by a factor of 2. Further increase of these truncation limits has no effect on the calculated vessel failure probabilities [14].

The only cladding flaws that contribute to the conditional probability of vessel failure must completely penetrate the clad layer (so that a crack tip resides in the potentially brittle ferritic steel), but no such flaws have been observed. However, owing to the limited amount of destructive examination on which this observation was based, the density of full-clad layer thickness flaws is modeled in FAVOR as 10 percent of the observed density of embedded flaws in the clad layer.

Additionally, the flaw model contains other features known to be accurate.

- The destructive examinations performed on Shoreham and PVRUF revealed that the overwhelming majority of weld flaws were lack of fusion defects and, so, occurred preferentially on the fusion line between the base metal and the weld metal. Consequently, in FAVOR, all weld flaws are modeled as occurring on the weld fusion line, which orients flaws in axial welds axially and flaws in circumferential welds circumferentially.

- The destructive examinations performed revealed that all cladding flaws are a lack of inter-run fusion defects and, so, occur parallel to the welding direction only. Consequently, in FAVOR, all clad flaws are modeled as occurring circumferentially in the vessel because the weld cladding was deposited circumferentially.

This combined use of empirical evidence, physical models, expert opinions, and conservative judgments allowed Simonen et al. to propose flaw distributions for use in FAVOR that are believed to be appropriate/conservative representations of the flaw population existing in PWRs in general (see Appendix A and Reference 40 for details). The overall conservatism associated with the Flaw Model led to its classification as red. Consequently, the NRC did not conduct sensitivity studies on the details of the flaw model. Nevertheless, it is informative to understand the characteristics of the flaws drawn from these distributions that contribute most significantly to the values of frequency of crack initiation (FCI) and TWCF estimated by FAVOR. Two general statements can be made regarding the flaws that contribute most significantly to the estimated TWCF values:

(1)   They are located close to the inner diameter surface of the vessel. The tensile thermal stresses produced by rapid cooling along the vessel inner diameter do not penetrate far into the wall thickness of the RPV. A natural consequence of this, which is illustrated in

Figure 4-3, is that the great majority of the cracks that are predicted to initiate and subsequently propagate through the vessel wall lie very close to the inner diameter surface. The information in Figure 4-3 indicates that almost all flaws that initiate lie less than 1/8 T from the vessel inner diameter. Since they are driven by the thermal stresses characteristic of cooldown transients, these observations hold true independent of embrittlement level.

(2)   They have a small through-wall dimension. This again occurs as a direct consequence of the fact that cooldown transients produce thermal stresses that, together with the pressure stresses, are only high enough to initiate cracks at locations close to the inner diameter of the vessel. Consequently, larger flaws (which would generally be considered more deleterious in a fracture evaluation than would small flaws) tend to not initiate very frequently because their crack tips lie too far away from the inner diameter surface and, so, are subjected to low tensile loads or even to compressive loads. Figure 4-4 and Figure 4-5 examine the effects of duration of irradiation exposure, flaw location (in plate or weld), and transient type on the flaw sizes that initiated fracture in the analyses. This information demonstrates that the combined effects of the duration of irradiation exposure and flaw location are small and entirely as expected because they correlate well with relative embrittlement levels. Transient type plays a minor role, with predominantly thermal transients, such as large pipe breaks, generally initiating fracture from smaller flaws, while transients that involve a significant pressure component (such as stuck-open valves that may later reclose) tend to initiate fracture from larger flaws. Nevertheless, the flaws that contribute to the estimated through-wall cracking frequency are small, having median depths ranging from 0.1 to 0.3 in.

In combination, these observations help to allay concerns that the flaw distributions sampled in FAVOR do not simulate enough flaws of large dimensions, or that the postulated future discovery of a large (previously undetected) flaw in service could invalidate the results of this study. Neither of these concerns is valid because, as a result of the dominant effects of thermal stresses in controlling crack driving force, large flaws do not play a dominant role in establishing the risk of RPV failure caused by PTS.

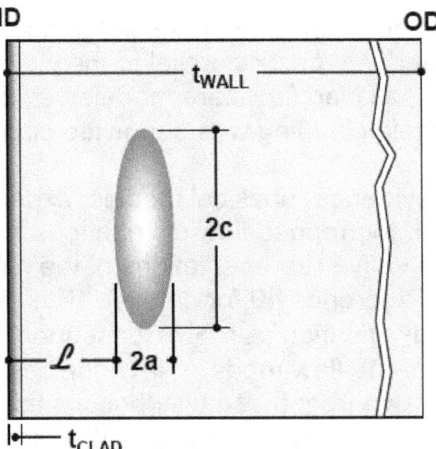

Figure 4-2 Flaw dimension and position descriptors adopted in FAVOR

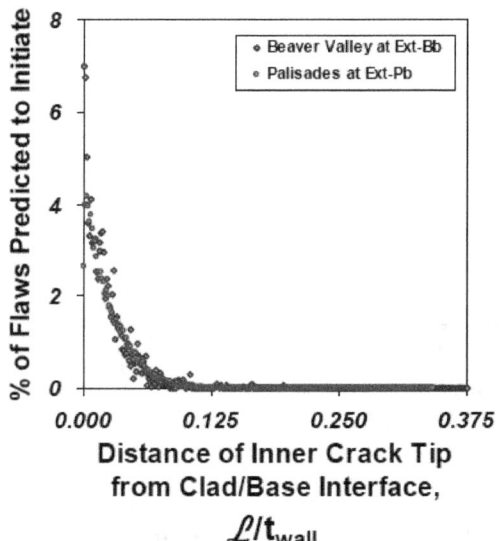

Figure 4-3 Distribution through-wall position of cracks that initiate

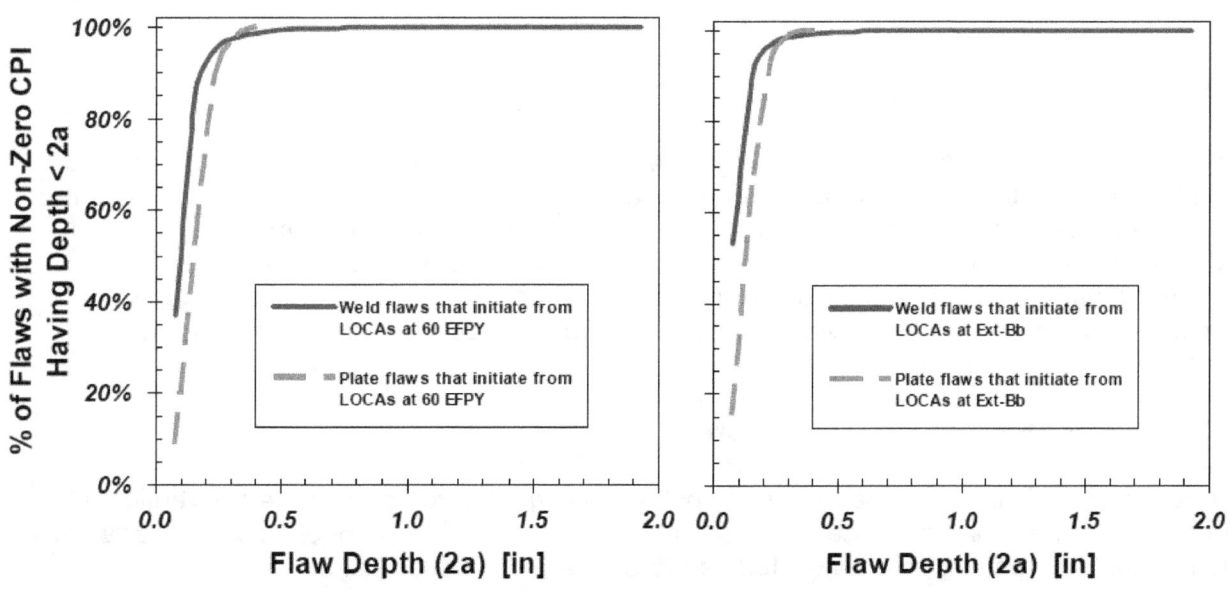

Figure 4-4 Flaw depths that contribute to crack initiation probability in Beaver Valley Unit 1 when subjected to medium- and large-diameter pipe break transients at two different embrittlement levels

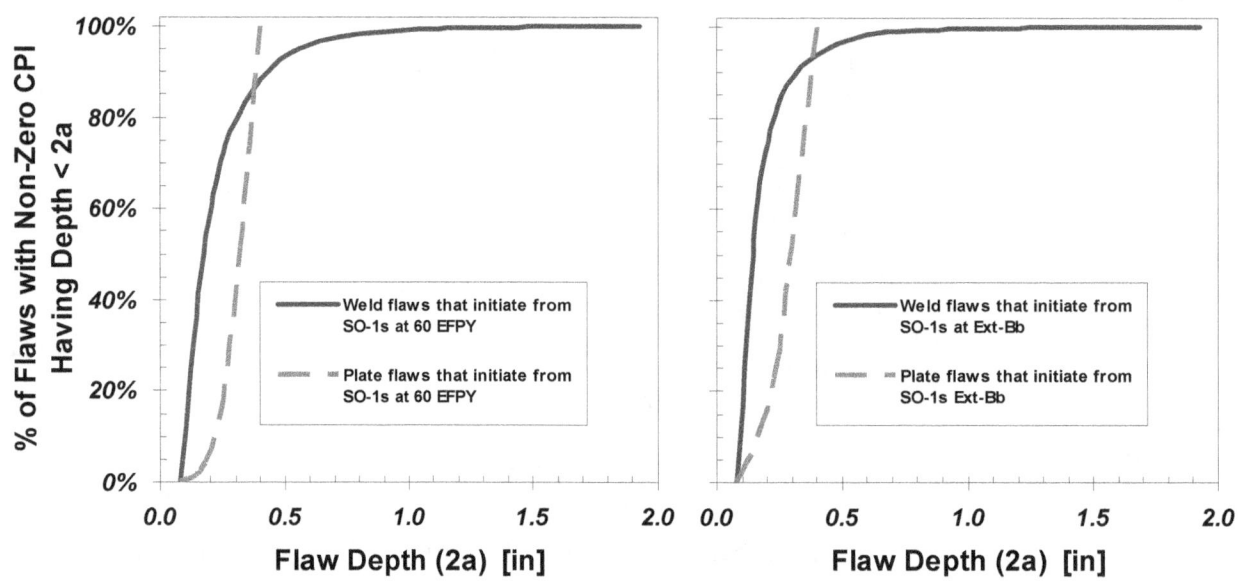

Figure 4-5 Flow depths that contribute to crack initiation probability in Beaver Valley Unit 1 when subjected to medium stuck-open valve transients at two different embrittlement levels

4.2 Crack Initiation Model

The crack initiation model, detailed in Figure 4-6, is essentially a comparison of the applied driving force to fracture ($K_{applied}$) and the material's resistance to crack initiation (K_{Ic}). The model K_{Ic} is itself made up of three major components—an unirradiated index temperature model, an index temperature shift model, and a toughness transition model. Section 4.2.1 describes sensitivity studies related to the fracture driving force model, which estimates $K_{applied}$, while Section 4.2.2 addresses sensitivity studies related to the model describing the material's resistance to crack initiation.

4.2.1 Applied Driving Force to Fracture

As illustrated in Figure 4-8, the model of applied driving force to fracture is a conventional linear elastic fracture mechanics model augmented by a warm prestress model. Various parts of the model, and the need for sensitivity studies thereon, are discussed below.

- FAVOR deterministically treats input variables describing the CTE$_{PLATE}$, CTE$_{CLAD}$, Vessel Diameter, Vessel Thickness, Elastic Modulus, and Poisson's Ratio. In all cases, the deterministic input value represents a best estimate. The uncertainty in these parameters is very small relative to many other variables in the model that have their uncertainties modeled explicitly (e.g., K_{Ic}, which exhibits uncertainty on the order of the mean value, and the uncertainty in the initiating event frequency can be several orders of magnitude). In the face of these much larger uncertainties it is not expected that the uncertainties on these input parameters influence the TWCF estimate in any significant manner. Consequently, no sensitivity studies are performed of these parameters.

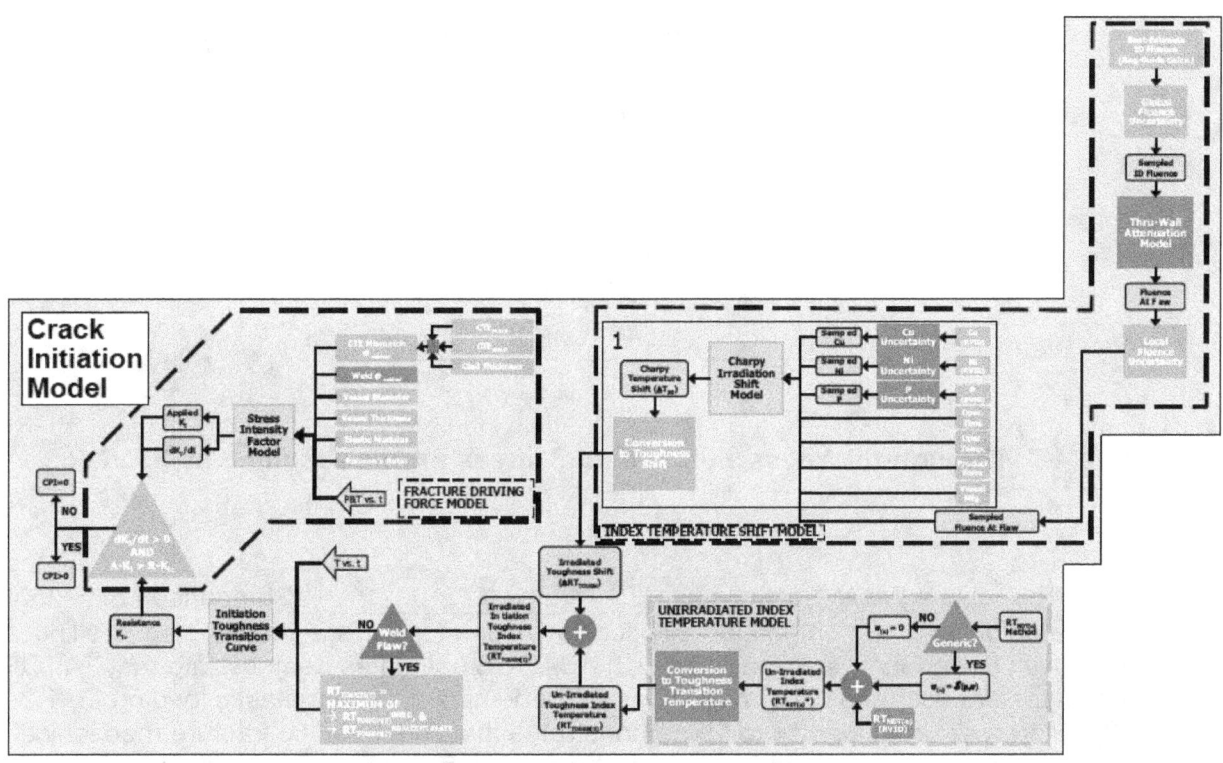

Figure 4-6 Schematic illustration of the crack initiation model used in the PTS reevaluation project

- FAVOR accurately models the effect of the ▮CTE MISMATCH RESIDUAL STRESS▮ on the estimated value of $K_{applied}$ because the $K_{applied}$ solution explicitly accounts for the CTE mismatch between the austenitic cladding and the ferritic base material. Therefore, no sensitivity study on CTE mismatch is performed.

- The ▮Weld Residual Stress Model▮ in FAVOR assumes that the stress distribution in Figure 4-7 quantifies the residual stresses produced by welding in both axial and circumferential welds. These residual stresses were estimated from measurements made of how the width of a radial slot cut in the longitudinal weld in a shell segment from an RPV changes with cut depth; these measurements were processed through a finite element analysis to determine the residual stress profile shown in Figure 4-7 [10]. FAVOR also assumes that this residual stress distribution is not relieved by cracking of the vessel, (i.e., the residual stresses to the right in the figure are applied equally irrespective of a/t). Since residual stresses have to be relieved if a crack were to develop through the weld in an RPV, a sensitivity study in which the weld residual stresses are retained in the crack initiation calculation, but are removed from the through-wall cracking calculation, is performed to assess the effect of this conservative assumption.

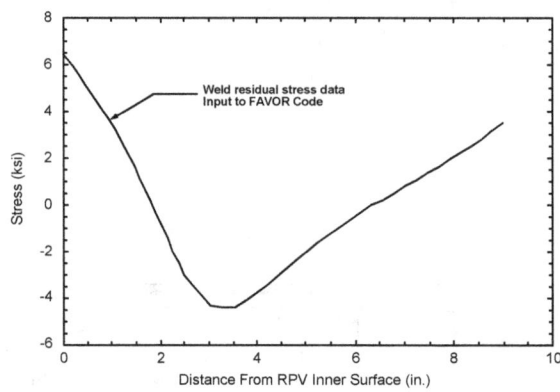

Figure 4-7 Through-thickness weld residual stress profile assumed by FAVOR
[56]

- FAVOR adopts a linear elastic Stress Intensity Factor Model. Comparison of FAVOR estimates of $K_{applied}$ for embedded elliptical flaws and for surface-breaking semi-elliptical flaws with closed form solutions and with ABAQUS estimates demonstrate that FAVOR estimates are accurate under conditions of predominantly linear elastic loading [9, 11, 12]. Moreover, an appendix to 19 demonstrates that linear elastic fracture mechanics (LEFM) are an accurate and appropriate theoretical framework for use in characterizing the resistance of thick-section nuclear RPVs to the structural integrity challenge posed by PTS loading. Consequently, sensitivity studies are not performed on the stress intensity factor model.

- FAVOR adopts a Warm Prestress Model (i.e., $dK_I/d_t>0$ and $K_{I(APPLIED)} > K_{Ic}$), as part of its linear elastic failure criteria. As detailed in Appendix B to 19 and in 23, warm prestress (WPS) is a physically understood and empirically demonstrated phenomenon. Of particular importance here, WPS has been demonstrated to be active in scale-model tests of nuclear RPVs subjected to PTS transients [4, 7]. For these reasons, a sensitivity study on the effects of the WPS model is not needed, and WPS is adopted as an integral part of the overall model. It should, however, be noted that adopting a WPS model can reduce the TWCF estimated for certain classes of transients. For example, the TWCF estimated for a primary-side pipe break will be significantly smaller when the effects of WPS are considered, while the TWCF estimated for a stuck-open valve that recloses later during the transient (thereby repressurizing the primary system) may not be affected by WPS at all. In plant analyses based on a complete set of transients (i.e., considering the potential for vessel failure from all potential PTS precursors), inclusion of WPS in the model reduces the estimated TWCF by between a factor of 2.5 and 3 [14]. Similar calculations reported by Meyer et al. indicate that WPS reduces TWCF by a factor of between 2 and 100 for individual transients [26]. The results reported by Dickson and by Meyer are not consistent because Dickson performed the analysis for a complete set of transients with the aim of assessing the aggregate PTS risk from all anticipated precursors, while Meyer focused on individual transients. The effect of WPS on individual transients reported by Dickson is within the range reported by Meyer.

4.2.2 Resistance to Crack Initiation

As illustrated in Figure 4-9, the model of the resistance of the material to crack initiation is itself made up of three major components—an unirradiated index temperature model, an index temperature shift model, and a toughness transition model. Sensitivity studies related to these three components are discussed, in turn, in each of the following sections.

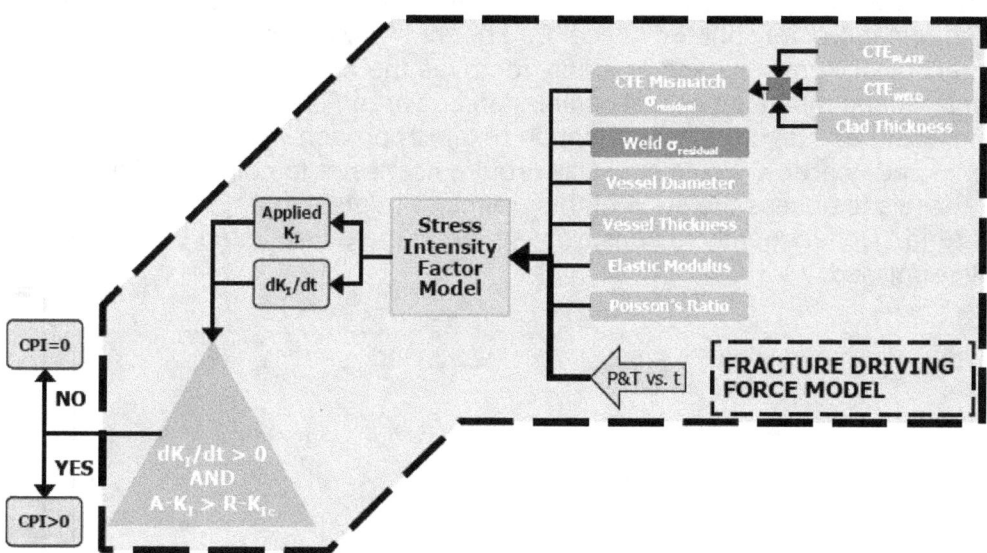

Figure 4-8 Schematic illustration of the model of the applied driving force to fracture used in the PTS reevaluation project

4.2.2.1 Unirradiated Index Temperature Model

The model adopted by FAVOR for the unirradiated index temperature recognizes that (1) the RT_{NDT} index temperature must be used in the definition of any revised PTS screening criteria so that plants can assess their reactors without the need to collect additional material data, and (2) the RT_{NDT} index temperature has an intentional conservative bias [2] that needs to be considered. In FAVOR, this

> **Conversion {of RT_{NDT}}**
> **to Toughness Transition**
> **Temperature**

is accomplished using a statistically defined conversion that quantifies the difference, and the uncertainty in the difference, between RT_{NDT} and a fracture toughness index temperature. While the conversion produces the correct result on average, it introduces a considerable nonphysical uncertainty that is an inescapable consequence of the inconsistency in the intentional conservative bias that is part of the ASME NB2331 RT_{NDT} definition. The result is that the uncertainty in values of the unirradiated toughness index temperature simulated by FAVOR greatly exceed those measured experimentally. For this reason, sensitivity studies examining alternative characterizations of RT_{NDT} and RT_{NDT} uncertainty are not conducted. FAVOR already overestimates RT_{NDT} uncertainty relative to any physically plausible or empirically justifiable characterization, and the only way to eliminate this conservatism is to eliminate RT_{NDT} from the model (which cannot be done because this project is constrained to assess the toughness properties of the RPV based only on information that is available currently).

With reference to Figure 4-9, it can also be noted that the overall **Unirradiated T$_{INDEX}$ Model** has been color coded to indicate that other plausible models exist that have roughly equal technical justification to the one adopted by FAVOR. This alternative would involve direct measurement of the index temperature using Master Curve technology, as detailed in ASTM E1921 and American Society of Mechanical Engineers (ASME) Code Case N-629. Such an approach is technically possible and arguably removes the considerable nonphysical uncertainty introduced by the statistical conversion between RT_{NDT} and the fracture toughness transition temperature. Nevertheless, adoption of this approach would also eliminate RT_{NDT} from the model and, thereby, would implicitly require licensees to collect additional material data on their RPV steels to assess the state of their reactors relative to any new proposed PTS screening criteria. Consequently, a sensitivity study on the unirradiated index temperature model is not performed.

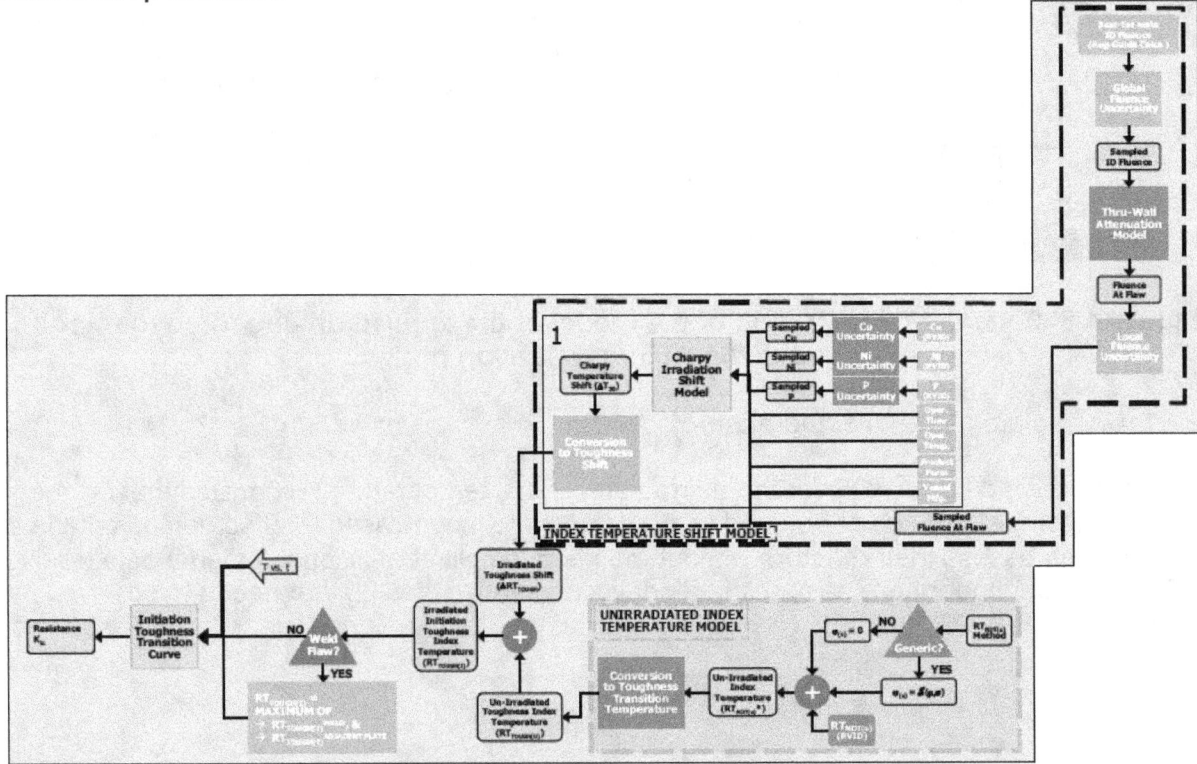

Figure 4-9 Schematic illustration of the model of the crack initiation toughness used in the PTS reevaluation project

4.2.2.2 Index Temperature Shift Model

The index temperature shift model includes a number of subcomponents that are discussed below in the order of their use in the FAVOR Index Temperature Shift Model:

- **ID Fluence**—The variation of fluence over the inner diameter of the vessel is estimated using modeling procedures based on the guidance provided in the NRC regulatory guide [34, "Calculational and Dosimetry Methods for Determining Pressure Vessel Neutron Fluence"]. Fluences calculated in this way are considered best estimates because they are based on the most up-to-date calculational procedures. Sensitivity studies involving inner diameter fluence are therefore not performed.

- **Thru-Wall Attenuation Model**—FAVOR adopts the RG 1.99, Revision 2, model of fluence attenuation through the thickness of the vessel [35]. This model assumes that the fluence (and thus the damage caused by irradiation) reduces exponentially as the through-wall distance from the inner radius of the RPV increases. The exponential coefficient adopted (−0.24) assumes that fluence attenuates at the same rate as displacements per atom, which is a conservative assumption. A recent review of attenuation models conducted under an EPRI contract [18] concluded that while the RG 1.99, Revision 2, attenuation model is widely regarded as conservative, no better alternative model exists at the current time. For this reason, no sensitivity study involving the attenuation model is performed.

- **Cu**, **Ni**, **P**—The mean values of copper (Cu), nickel (Ni), and phosphorus (P) used by FAVOR are the "best-estimate" values docketed by the licensees and recorded in the RVID2 database [31]. The statistical distributions assumed to exist around these mean values were derived from all data available regarding chemical constituent variability in nuclear-grade RPV steels and weldments (see Reference 19 for further explanation). Consequently, these distributions overestimate (sometimes significantly so) the degree of uncertainty in Cu, Ni, and P relative to what actually exists in a particular plate, weld, or forging. While sensitivity studies could be motivated on such a basis, and indeed might provide valuable insights about the PTS risk at particular plants, the results of such studies would not provide useful information about developing a PTS screening criteria applicable to the entire population of operating PWRs. For this reason, sensitivity studies on Cu, Ni, and P are not performed.

- **Time**, **Prod. Form**, **Mfg.**—Each FAVOR analysis is performed at a particular value of reactor operating time defined in terms of effective full-power years, or EFPYs. As such, the variable time is known without error. Records may also be used to define without question the product form (weld, plate, or forging), as well as the company responsible for manufacturing each vessel. Sensitivity studies on these variables are therefore not needed.

- **Temperature**—Since damage of the RPV steels by neutron irradiation exposure is a phenomenon that occurs on a time scale of years to decades, transients in the operating temperature of the reactor are of no significance to the predictions of this model. The temperature value needed is that in the beltline region of the reactor when it is operating under full-power conditions. This temperature can be established from operational records. For these reasons, sensitivity studies on the operating temperature are not needed.

- **Charpy Irradiation Shift Model**—This model, also called an embrittlement trend curve, relates compositional and exposure variables to the amount by which irradiation shifts the Charpy V-notch (CVN) transition temperature curve to higher temperatures. FAVOR adopts a model developed by Eason under an NRC research contract [17]. Since that time, the American Society for Testing and Materials adopted a similar, but not identical, embrittlement trend curve in the E900-02 standard [3]. A sensitivity study will be performed to assess the effect of adopting the ASTM embrittlement trend curve rather than the model proposed by Eason.

- **Conversion to Toughness Shift**—This model converts the shift in the CVN transition temperature into a shift in the fracture toughness transition temperature. The relationship between these two transition temperatures has been shown to be consistent over a wide range of RPV materials, although it is dependent on product form [42, 22]. While there can appear to be considerable uncertainty in this relationship, it has been demonstrated that this uncertainty derives principally from an inability of small data sets to resolve accurately the transition temperatures [43, 22]. For these reasons no sensitivity studies are performed.

4.2.2.3 Toughness Transition Model

The toughness transition model adopted by FAVOR (i.e., the variation of K_{lc} with temperature) was fit to all available LEFM-valid fracture toughness data for domestic RPV steels [56]. The model is not, however, a purely empirical fit since the dispersion coefficient (c) is held fixed at 4, a value motivated by a physical understanding of cleavage fracture as expressed by Wallin's Master Curve (see all of the references by Wallin).

The color-coding of the **Toughness Transition Model**, signifies that alternative models having roughly equal technical justification are available. This alternative model is Wallin's Master Curve. The NRC did not view the adoption of Wallin's Master Curve for the Toughness Transition Model used by FAVOR as appropriate because (1) the Master Curve provides the temperature dependency of the elastic-plastic fracture toughness parameter K_{Jc}, and (2) the Master Curve incorporates explicitly the statistical size dependency of fracture toughness values. Both of these features of the Master Curve are inconsistent with the framework of FAVOR, which is a linear elastic fracture mechanics code. For these reasons, and because adapting FAVOR to include an elastic-plastic fracture toughness model represents a major change to the structure of the code, no sensitivity studies are performed.

4.2.2.4 Other Features

For weld flaws (which are all fusion line flaws), the FAVOR crack initiation model assumes that the fracture toughness transition reference temperature that characterizes the material at the crack tip is the maximum of the fracture toughness transition reference temperature of the plate material that exists on one side of the flaw and the weld material that exists on the other. This model has been classified as **Green (i.e., accurate)** because a crack, if it propagates, will tend to do so through the most brittle material available. For this reason, no sensitivity studies of this model are performed.

4.3 Through-Wall Cracking Model

As illustrated in Figure 4-10, the model of through-wall cracking adopted by FAVOR compares the applied driving force to fracture (A-K_l) and the material's resistance to further cracking (R-K_l) to estimate the conditional probability of through-wall cracking (which is defined as vessel failure (CPF)). The through-wall cracking model is executed 100 times by FAVOR for each time step in the transient at which the crack initiation model estimates a nonzero conditional probability of crack initiation (CPI). Each of these 100 trials constitutes a single, deterministic, crack-arrest calculation, and the fraction of these 100 trials for which through-wall cracking is predicted provides an estimate of the percentage of CPI that is manifested in CPF.

The through-wall cracking model is composed of the following parts:

- a fracture-driving-force model

- a fracture resistance model, which is made up of the following parts:

  — the assumption that upon crack initiation the crack immediately grows along the inner diameter surface of the vessel to a length that greatly exceeds its through-wall depth

  — a model that describes the gradient of weld properties through the vessel wall thickness

  — a crack-arrest model

  — a ductile-tearing model

Section 4.3.1 describes the sensitivity studies related to the fracture-driving-force model, while Section 4.2.2 addresses sensitivity studies related to the four different parts of the fracture resistance model.

4.3.1 Fracture Driving Force Model for Through-Wall Cracking

Similar to the crack initiation model, a conventional linear elastic fracture mechanics approach is taken to characterize the applied driving force to fracture (see Figure 4-11). Most details of this model are identical to those already discussed in Section 4.2.1 and thus are not repeated. With respect to the **Stress Intensity Factor Model**, comparison of FAVOR estimates of $K_{applied}$ for axial and circumferential flaws of infinite length with closed form solutions and with the ABAQUS estimates demonstrate that FAVOR estimates are accurate under conditions of predominantly linear elastic loading [5]. Moreover, an appendix to 19 demonstrates that LEFM is an accurate and appropriate theoretical framework for use in characterizing the resistance of thick-section nuclear RPVs to the structural integrity challenge posed by PTS loading.

4.3.2 Fracture Resistance Model for Through-Wall Cracking

As discussed earlier, the fracture resistance model for through-wall cracking is made up of the following parts:

- the assumption that upon crack initiation the crack immediately grows along the inner diameter surface of the vessel to a length that greatly exceeds its through-wall depth

- a model that describes the gradient of weld properties through the vessel wall thickness

- a crack arrest model

- a ductile tearing model

Sensitivity studies relating to each part of the fracture resistance model for through-wall cracking are discussed in the following sections.

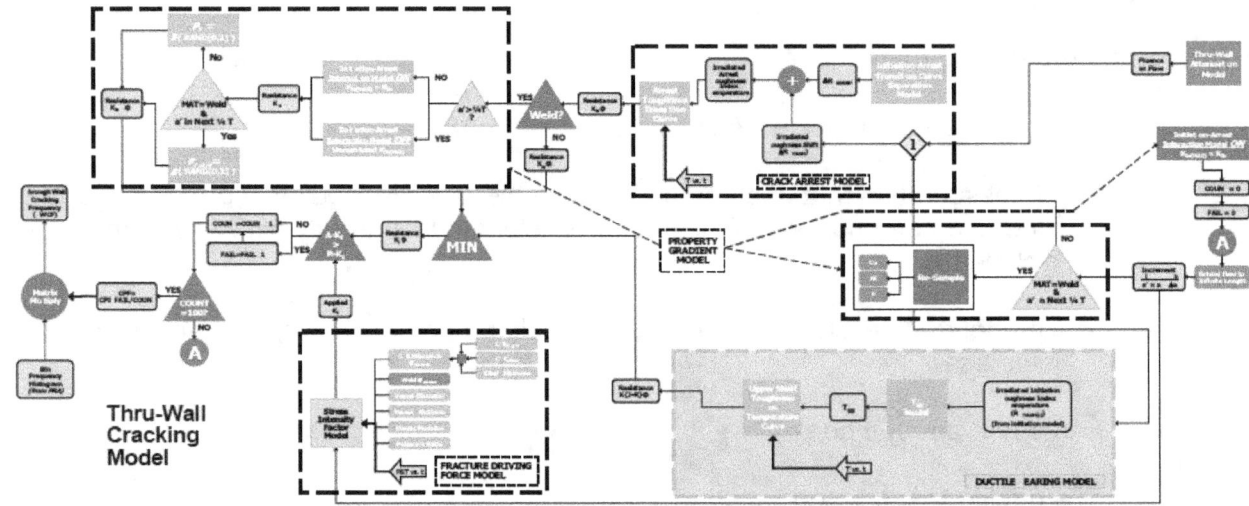

Thru-Wall
Cracking
Model

Figure 4-10 Model of through-wall cracking adopted by FAVOR

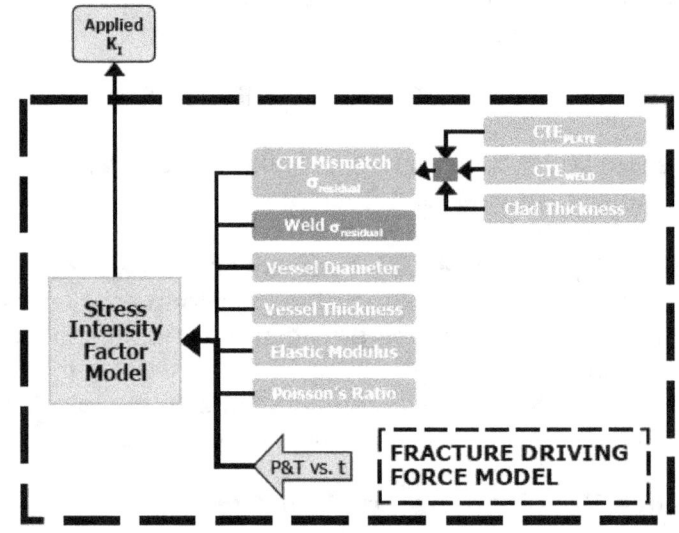

Figure 4-11 Crack driving force model for through-wall cracking

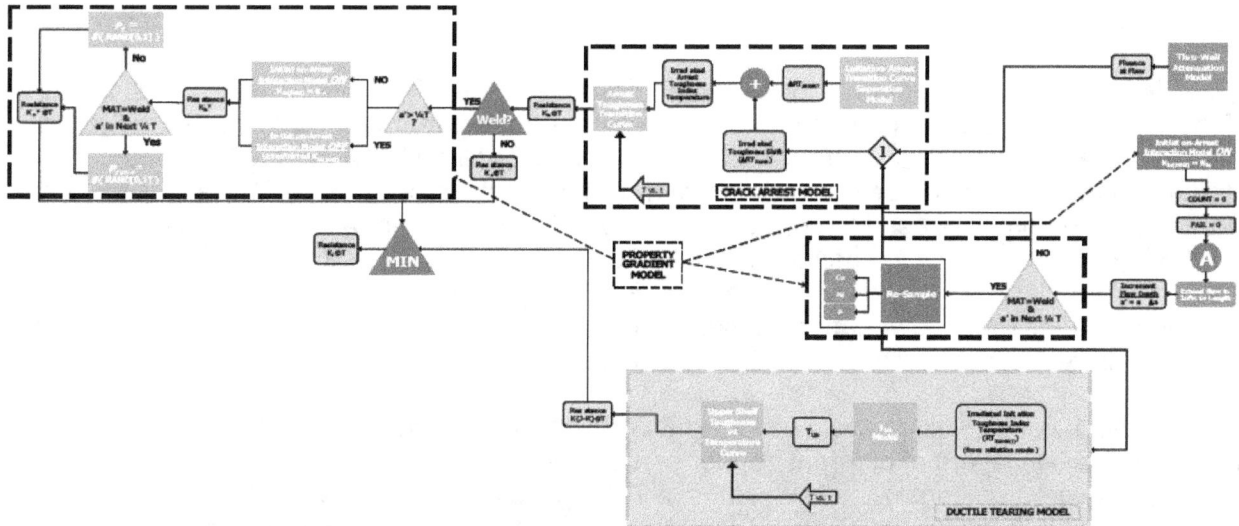

Figure 4-12 Fracture resistance model for through-wall cracking

4.3.2.1 Infinite Length Flaw Assumption

Whenever the crack initiation model estimates a nonzero conditional probability of crack initiation, FAVOR Extends to Infinite Length the surface flaw or buried elliptical flaw that caused this initiation. The assumption that flaws extend in length (axially or circumferentially) by considerable distances before they extend significantly in depth (through the vessel wall thickness) is supported by thermal shock experiments performed on scaled RPVs [7]. However, use of this assumption when trying to predict the fracture of a real vessel is believed to be conservative for the following reasons:

- The total propagation length of axial flaws is expected to be limited to a single shell course in most cases (slightly over 8 ft long) because at this point the crack would propagate first into circumferential welds and, secondly, into a plate or an axial weld. These other materials are not likely to be as embrittled as the region where the crack started, and so are likely to arrest the crack. In any event, the active core (approximately 12 ft in length) is the only region over which significant embrittlement can occur, so crack arrest would certainly occur outside of this region.

- The total propagation length of circumferential flaws is expected to be limited by the periodic circumferential variations of fluence that occur because of varying distances between the fuel and the vessel wall. At least every 45 degrees (e.g., approximately every 5.25 ft in a 13-ft diameter vessel), the fluence in the core region goes through a minima, suggesting an increased likelihood of crack arrest at these less embrittled locations.

While the treatment of flaw length adopted by FAVOR is conservative, as detailed above, it is not believed to be excessively so because the flaw depth (on the order of 0.01 to 1 in.) is very small compared to these flaw lengths (100 to 160 in.). Ample evidence in the fracture mechanics literature demonstrates that as the length-to-depth ratio for a flaw becomes large (i.e., above 30), the correct $K_{applied}$ solution is closely approximated by the infinite length assumption. For this reason no sensitivity studies are performed on this topic.

4.3.2.2 Weld Property Gradient Model

In welds, a gradient of properties through the thickness of the RPV is expected to exist because of through-wall changes in copper content. These copper content changes arise from the fact that, because of the large volume of weld metal needed to fill an RPV weld, oftentimes manufacturers needed to use weld wire from multiple spools to completely fill the groove. Lack of control of the process used to copperplate the weld wires (a step taken for corrosion control) resulted in wide variability in coating thickness from spool to spool, which is manifested in measurable variations in reported copper content through the RPV wall thickness. These copper variations produce variations in sensitivity to irradiation embrittlement and consequent variations in resistance to fracture though the vessel wall.

Simple calculations involving weld coil weight and weld volume suggest that axial welds are likely to have 2–3 layers of different copper content, while circumferential welds (owing to their considerably greater length) are likely to have 7–8 layers [19]. However, FAVOR adopts a simpler model that **All Welds Have 4 Layers** and, so, chemical composition (Cu, Ni, P)[§] is resampled when the crack propagates into a new weld layer. A sensitivity study on this model is performed in which all weld layers are removed from the model.

Another feature of the weld property gradient model is the Initiation-Arrest Interaction Model. Physical arguments suggest that within the same material the fact that a crack has initiated (at, for example, $K_{Ic}*$) suggests that the crack arrest toughness at the temperature of crack initiation cannot exceed $K_{Ic}*$, or arrest would have occurred immediately and the initiation event would not have been successful [19]. However, no expectations on K_{Ia} values exist for a material in which crack initiation has not yet occurred. For this reason, once Cu, Ni, and P are resampled, all restrictions on K_{Ia} based on the K_{Ic} observed in the initiating weld layer are relaxed. This Initiation-Arrest Interaction Model is premised on a physical understanding of the processes responsible for cleavage fracture and, therefore, no sensitivity studies are planned.

4.3.2.3 Crack Arrest Model

19 discusses in detail the model of cleavage crack arrest toughness adopted by FAVOR and illustrated in Figure 4-12. This model includes an Initiation-Arrest Transition Curve Separation Model that varies depending on the crack initiation toughness index temperature, and an Arrest Toughness Transition Curve that is indexed based on fracture toughness data rather than the correlative index temperature RT_{NDT}. In combination, these treatments model available K_{Ia} data more realistically than the only available alternative model (i.e., the ASME K_{Ia} curve). Furthermore, the material-dependent separation of the K_{Ia} curve from the K_{Ic} curve conforms to a physical understanding of the mechanics of cleavage fracture in a way that the ASME K_{Ia} model does not. Thus, while alternatives to the crack arrest model adopted by FAVOR exist, it cannot be said that they have "roughly equal technical justification" to the FAVOR model. In fact, the alternative model is not as well justified, either empirically or physically, as the FAVOR model. For this reason no sensitivity studies based on these models are performed.

[§] See Section 4.2.2.2 for a description of sensitivity studies related to chemical composition.

4.3.2.4 Ductile Tearing Model

The model of ductile tearing adopted by FAVOR and illustrated in Figure 4-12 is discussed in detail elsewhere (see Reference 21 for model description and Reference 19 and Reference 56 for a discussion of its use in FAVOR). This model includes a T_{US} Model that estimates an index temperature for upper-shelf fracture toughness from the index temperature for transition fracture toughness, and is used along with an Upper Shelf Toughness vs. Temperature Curve to estimate the resistance to ductile tearing (*J-R*) at temperatures of interest. The detailed discussion of these models (see both Reference 19 and Reference 21) demonstrates that they apply, and can be expected to apply, to all material conditions of interest in RPV applications.

Alternative models of ductile tearing resistance in ferritic steels to that illustrated in Figure 4-12 exist [16, 36, 54], and it is for this reason that the overall **Ductile Tearing Model** is classified as orange in Figure 4-12. These alternatives, which are all based on correlations with CVN upper-shelf energy, are much more commonplace than that adopted in FAVOR, which does not rely on Charpy correlations in any way and features an explicit treatment of the uncertainty in upper-shelf toughness data. The model in FAVOR estimates the upper-shelf toughness properties directly from the fracture toughness transition temperature, a relationship motivated by both trends in fracture toughness data and by physical considerations. The NRC implemented the new model in FAVOR instead of the more conventional approaches based on correlation with CVN properties becasue of the notoriously poor correlation coefficients exhibited by these other approaches ($R^2 < 0.5$; see discussion in Chapter 9 of Reference 19). This lack of correlation suggests that different physical mechanisms underlie energy absorption in the CVN test versus the resistance-to-ductile crack initiation and propagation from a preexisting defect. Additionally, these low correlation coefficients engender little confidence in the reliability of calculations made on the basis of such relationships. However, because approaches to upper-shelf toughness estimation based on correlation with CVN properties are much more commonplace, a sensitivity study on the **Ductile Tearing Model** is planned.

4.4 Summary of Planned Sensitivity Studies

The following list summarizes the sensitivity studies proposed in the preceding sections:

- Welding residual stresses are currently assumed to not be affected by through-wall crack propagation. A sensitivity study is performed in which the residual stresses are set to zero once the crack initiates.

- FAVOR currently adopts an embrittlement shift model that differs from that recommended by ASTM. A sensitivity study is performed in which the ASTM model is adopted.

- FAVOR resamples chemical composition variables at the 1/4 T, 1/2 T, and 3/4 T locations for welds in a through-wall propagation analysis to simulate the effect of varying copper content on different weld wire spools. A sensitivity study is performed in which the chemistry is not resampled.

- FAVOR adopts a new toughness-based upper-shelf model instead of more conventional approaches based on correlation with CVN properties. A sensitivity study is performed in which a CVN-based upper-shelf model is adopted.

The material conditions and TH transient sets used for each sensitivity study, and the rationale for making these selections, include the following:

- Material Condition—Results of the baseline PTS analyses [15] and previous results [24] all demonstrate that the embrittlement properties of materials that can be associated with an axial crack play the dominant role in establishing the risk associated with a PTS event. For this reason, the NRC selected a plant in which the axial welds are the most embrittled material (Palisades) and a plant in which the plates are the most embrittled material (Beaver Valley) for use in sensitivity studies. Two different embrittlement conditions are analyzed for each plant, (1) the embrittlement expected to occur at the end of the plant's original 40-year operating license (32 EFPYs), and (2) a much higher embrittlement level expected to coincide (approximately) with a yearly through-wall cracking frequency of 1×10^{-6} events/year. Table 4-2 summarizes these different embrittlement conditions. Each sensitivity study listed above is analyzed for all four of the conditions listed in Table 4-2.

- Thermal Hydraulic Transient Subset—The subset of transients from the base case analyses [24] that contribute the most to the TWCF are used in these analyses. Figure 4-13 illustrates this subset for the Palisades and Beaver Valley reactors, while Table 4-3 and Table 4-4 provide a detailed description of these transients.

Table 4-2 Plant and Embrittlement Conditions Used in Sensitivity Studies

Plant Type	Embrittlement Level	
	EOL	Near Proposed RVFF limit of 10^{-6} events/year
Axial Weld Limited	Palisades at 32 EFPYs	Palisades at Ext-B
Plate Limited	Beaver at 32 EFPYs	Beaver at Ext-B

Table 4-3 Transients Used in Sensitivity Studies for Palisades

Case	Class	System Failure	Operator Action	HZP
19	SO-2	Reactor trip with 1 stuck-open ADV on SG-A.	None. Operator does not throttle HPI.	Yes
40	LOCA	40.64-cm (16-in.) hot-leg break. Containment sump recirculation included in the analysis.	None. Operator does not throttle HPI.	No
52	SO-2	Reactor trip with 1 stuck-open ADV on SG-A. Failure of both MSIVs (SG-A and SG-B) to close.	Operator does not isolate AFW on affected SG. Normal AFW flow assumed (200 gpm). Operator does not throttle HPI.	Yes
54	MSLB	Main steamline break with failure of both MSIVs to close. Break assumed to be inside containment causing containment spray actuation.	Operator does not isolate AFW on affected SG. Operator does not throttle HPI.	No
55	SO-2	Turbine/reactor trip with 2 stuck-open ADVs on SG-A combined with controller failure resulting in the flow from two AFW pumps into affected steam generator.	Operator starts second AFW pump.	No
58	LOCA	10.16-cm (4-in.) cold-leg break. Winter conditions assumed (HPI and LPI injection temp = 40 °F, Accumulator temp = 60 °F)	None. Operator does not throttle HPI.	No

Case	Class	System Failure	Operator Action	HZP
59	LOCA	10.16-cm (4-in.) cold-leg break. Summer conditions assumed (HPI and LPI injection temp = 100 °F, Accumulator temp = 90 °F)	None. Operator does not throttle HPI.	No
60	LOCA	5.08-cm (2-in.) surge line break. Winter conditions assumed (HPI and LPI injection temp = 40 °F, Accumulator temp = 60 °F)	None. Operator does not throttle HPI.	No
62	LOCA	20.32-cm (8-in.) cold-leg break. Winter conditions assumed (HPI and LPI injection temp = 40 °F, Accumulator temp = 60 °F)	None. Operator does not throttle HPI.	No
63	LOCA	14.37-cm (5.656-in.) cold-leg break. Winter conditions assumed (HPI and LPI injection temp = 40° F, Accumulator temp = 60 °F)	None. Operator does not throttle HPI.	No
64	LOCA	10.16-cm (4-in.) surge line break. Summer conditions assumed (HPI and LPI injection temp = 100 °F, Accumulator temp = 90 °F)	None. Operator does not throttle HPI.	No
65	SO-1	One stuck-open pressurizer SRV that recloses at 6000 s after initiation. Containment spray is assumed not to actuate.	None. Operator does not throttle HPI.	Yes

Table 4-4 Transients Used in Sensitivity Studies for Beaver Valley Unit 1

Case	Class	System Failure	Operator Action	HZP
7	LOCA	2.54-cm (8-in.) surge line break	None.	No
9	LOCA	2.54-cm (16-in.) hot-leg break	None.	No
56	LOCA	10.16-cm (4.0-in.) surge line break		
60	SO-1	Reactor/turbine trip w/one stuck-open pressurizer SRV which recloses at 6000 s.	None.	No
96	SO-1	Reactor/turbine trip w/one stuck-open pressurizer SRV which recloses at 6000 s.	Operator controls HHSI 10 minutes after allowed.	No
97	SO-1	Reactor/turbine trip w/one stuck-open pressurizer SRV which recloses at 3000 s.	None.	Yes
101	SO-1	Reactor/turbine trip w/one stuck-open pressurizer SRV which recloses at 3000 s.	Operator controls HHSI 10 minutes after allowed.	Yes
102	MSLB	Main steamline break with AFW continuing to feed affected generator for 30 minutes.	Operator controls HHSI 30 minutes after allowed. Break is assumed to occur inside containment so that the operator trips the RCPs as a result of adverse containment conditions.	No
103	MSLB	Main steamline break with AFW continuing to feed affected generator for 30 minutes.	Operator controls HHSI 30 minutes after allowed. Break is assumed to occur inside containment so that the operator trips the RCPs as a result of adverse containment conditions.	Yes
104	MSLB	Main steamline break with AFW continuing to feed affected generator for 30 minutes.	Operator controls HHSI 60 minutes after allowed. Break is assumed to occur inside containment so that the operator trips the RCPs as a result of adverse containment conditions.	No
105	MSLB	Main steamline break with AFW continuing to feed affected generator for 30 minutes.	Operator controls HHSI 60 minutes after allowed. Break is assumed to occur inside containment so that the operator trips the RCPs as a result of adverse containment conditions.	Yes
108	MSLB	Small steamline break (simulated by sticking open all SG-A SRVs) with AFW continuing to feed affected generator for 30 minutes.	Operator controls HHSI 30 minutes after allowed.	Yes

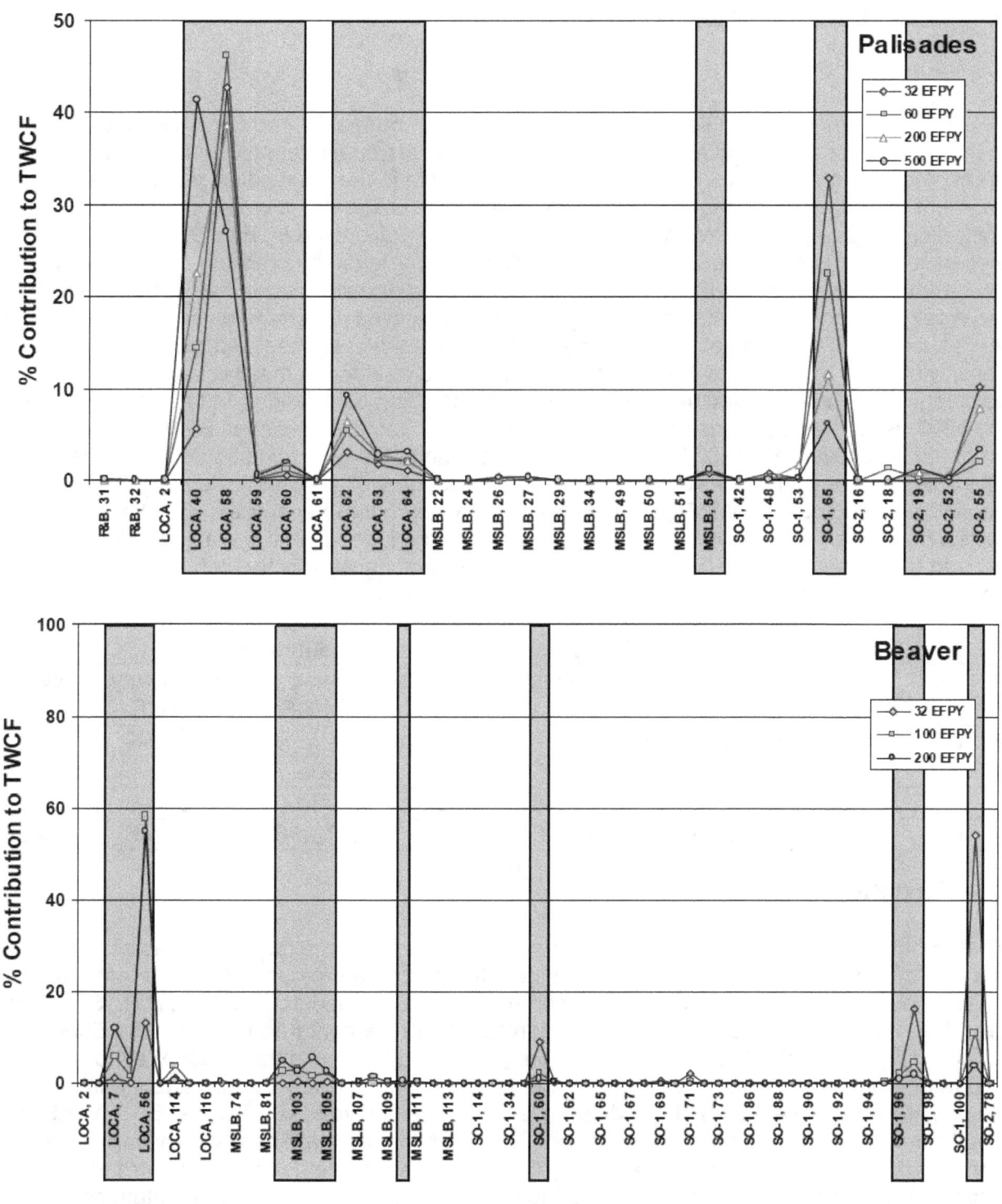

Figure 4-13 Identification of transients (shaded boxes) used for Palisades and Beaver Valley sensitivity studies

4.5 Sensitivity Study Results

4.5.1 Residual Stresses

FAVOR assumes that a single distribution quantifies, for both axial and circumferential welds, the residual stresses produced by welding [56]. These residual stresses were estimated from measurements made of how the width of a radial slot cut in the longitudinal weld in a shell segment from an RPV changes with cut depth; these measurements were processed through a finite element analysis to determine the residual stress profile used by FAVOR [10]. FAVOR also assumes that this residual stress distribution is not relieved by cracking of the vessel (i.e., the residual stresses in the figure to the right are applied equally irrespective of the a/t ratio of the crack). Since residual stresses would have to be relieved if a crack were to develop through the weld in an RPV, the effect of this conservative assumption is assessed by performing a sensitivity study in which the weld residual stresses are retained in the crack initiation calculation, but are removed from the through-wall cracking calculation. In this sensitivity study, the NRC performed analyses of both the Beaver Valley and the Palisades RPVs at two embrittlement levels each (32 EFPYs and the Ext-B embrittlement conditions; see Table 4-5). The effect of relieving the residual stresses in the through-wall cracking calculations was entirely negligible, reducing the TWCF values by less than 1 percent (on average). This limited sensitivity of the TWCF values on residual stresses occurs because the crack driving force caused by the residual stress is very small relative to that caused by the combination of thermal and pressure loading.

Table 4-5 Results of Residual Stress Sensitivity Study

Plant	EFPYs	Base FCI	Sensitivity FCI	FCI Ratio	Base TWCF	Sensitivity TWCF	TWCF Ratio
BV	32	1.56e-7	1.56e-7	1.00	1.40e-9	1.55e-9	1.10
BV	200	8.48e-6	8.46e-6	0.99	2.80e-7	2.67e-7	0.95
Pal	32	9.67e-8	9.75e-8	1.00	9.98e-9	9.56e-9	0.96
Pal	500	5.83e-6	5.73e-6	0.98	1.85e-6	1.77e-6	0.96

4.5.2 Embrittlement Shift Model

The embrittlement shift model relates compositional and neutron exposure variables to the amount by which irradiation shifts the CVN transition temperature curve to higher temperatures. FAVOR adopts a model developed by Eason under an NRC research contract [17]. Since that time ASTM has adopted a similar, but not identical, embrittlement trend curve in the E900-02 standard [3]. A sensitivity study was therefore performed to assess the effect of adopting the ASTM embrittlement trend curve rather than the model proposed by Eason (again analyzing Beaver Valley and Palisades at two different embrittlement levels; see Table 4-6). The ASTM E900-02 embrittlement shift model produces TWCF estimates that are systematically lower (i.e., approximately 1/3 of those estimated using the Eason shift model). Activity is currently underway within ASTM Committee E10.02 to revise the E900 model. Representatives of both the industry and the NRC are involved in this Code committee work, and it is expected that the committee will publish a revised model that incorporates features of both the current Eason and ASTM E900-02 relationships. Thus, for the purposes this report, the Agency has continued to use the Eason correlation and accepted this approach as conservative relative to the approach adopted by an international consensus body. When a consensus emerges from the ASTM

E10.02 Code committee process it will be a simple matter to assess the effect of the new embrittlement shift model on the TWCF values reported in this report.

4.5.3 Chemical Composition Resampling for Welds

In welds, a gradient of properties through the thickness of the RPV is expected to exist because of through-wall changes in copper content. These copper content changes arise from the fact that, because of the large volume of weld metal needed to fill an RPV weld, oftentimes manufacturers used weld wire from multiple weld wire spools to completely fill the groove. Lack of control of the process used to copperplate the weld wires (a step taken for corrosion control) resulted in wide variability in copper coating thickness from spool to spool, which is manifested in measurable variations in copper content through the RPV wall thickness. These copper variations produce variations in sensitivity to irradiation embrittlement, and consequent variations in resistance to fracture though the vessel wall.

FAVOR adopts a weld composition gradient model in which the copper content is resampled in a through-wall cracking calculation every time the crack passes the 1/4 thickness, the 1/2 thickness, and the 3/4 thickness locations in the vessel wall. A four-weld layer model was developed based on considerations of the volume of weld metal needed to fill an RPV weld(see Reference 19) To assess the effect of this model on the TWCF, a sensitivity study was performed in which the copper resampling feature in FAVOR was deactivated. Again, the sensitivity study included analysis of Beaver Valley and Palisades at two different embrittlement levels (see Table 4-7). The results of this study show that turning off the FAVOR 4-weld layer model increases the estimated TWCF by a small amount (i.e., a factor of 2.5 on average).

Table 4-6 Results of Embrittlement Shift Model Sensitivity Study

Plant	EFPYs	Base FCI	Sensitivity FCI	FCI Ratio	Base TWCF	Sensitivity TWCF	TWCF Ratio
BV	32	1.56e-7	4.26e-8	0.273	1.40e-9	3.68e-10	0.263
BV	200	8.48e-6	2.75e-6	0.324	2.80e-7	6.05e-8	0.216
Pal	32	9.67e-8	2.77e-8	0.286	9.98e-9	1.56e-9	0.156
Pal	500	5.83e-6	4.68e-6	0.803	1.85e-6	1.36e-6	0.735

Table 4-7 Results of Chemistry Resampling Sensitivity Study

Plant	EFPYs	Base FCI	Sensitivity FCI	FCI Ratio	Base TWCF	Sensitivity TWCF	TWCF Ratio
BV	32	1.56e-7	1.56e-7	1.00	1.40e-9	2.09e-9	1.49
BV	200	8.48e-6	8.48e-6	1.00	2.80e-7	3.65e-7	1.30
Pal	32	9.67e-8	9.75e-8	1.01	9.98e-9	5.39e-8	5.40
Pal	500	5.83e-6	5.83e-6	1.00	1.85e-6	2.99E-6	1.62

4.5.4 Upper-Shelf Toughness Model

In FAVOR Version 3.1, upper-shelf fracture toughness values (J_{Ic}, J-R) were estimated through correlations with CVN energy. These empirical relationships had very low correlation coefficients and high scatter, reflecting the different underlying physical processes that control

Charpy energy and fracture toughness on the upper shelf. Comments from the peer review group (see comment #40, Appendix B to Reference 20) questioned the appropriateness of this approach. After reviewing the existing FAVOR model and other available alternatives, the staff adopted and implemented a new upper-shelf model in FAVOR Version 04.1 to address this concern. This new model does not rely on Charpy correlations in any way and features an explicit treatment of the uncertainty in upper-shelf toughness (both the ductile initiation toughness as measured by J_{Ic} and the resistance to further crack extension as measured by J-R). Additionally, the new model links transition toughness and upper-shelf toughness properties, a relationship motivated by trends in fracture toughness data and physical considerations. The FAVOR Version 03.1 models did not have these features.

As demonstrated by the results in Table 4-8, the new upper-shelf model does not change the TWCF values in any substantive way. On average, the TWCF values estimated using the new model are approximately 5 percent lower than the values estimated by the correlative approaches used in FAVOR Version 03.1. However, the linkage between transition toughness and upper-shelf toughness properties in the new model has eliminated FAVOR predictions of physically implausible results (e.g., predicting that flaws in a particular axial weld, such as Axial Weld A, of the RPV beltline contribute more to the TWCF than do flaws in another axial weld, such as Axial Weld B, even though the toughness of Axial Weld A exceeds that of Axial Weld B.

Table 4-8 Results of Upper-Shelf Toughness Model Sensitivity Study

Plant	EFPYs	TWCF with CVN-Based Upper--Shelf Model (FAVOR 3.1)	TWCF with Toughness-Based Upper Shelf Model (FAVOR 4.1)	TWCF Ratio (Toughness Model / CVN Model)
Oconee	32	2.39E-11	2.30E-11	0.96
	60	8.39E-11	6.47E-11	0.77
	Ext-Oa	2.44E-09	1.30E-09	0.53
	Ext-Ob	1.82E-08	1.16E-08	0.64
Beaver Valley	32	5.31E-10	8.89E-10	1.67
	60	5.30E-09	4.84E-09	0.91
	Ext-Ba	2.68E-08	2.02E-08	0.75
	Ext-Bb	4.06E-07	3.00E-07	0.74
Palisades	32	6.20E-09	4.90E-09	0.79
	60	1.51E-08	1.55E-08	1.03
	Ext-Pa	1.42E-07	1.88E-07	1.32
	Ext-Pb	1.14E-06	1.26E-06	1.11
			Average	0.94

5. SENSITIVITY STUDIES PERFORMED TO PROVIDE CONFIDENCE IN THE APPLICABILITY OF TWCF RESULTS TO PWRS IN GENERAL

The NRC performed the following sensitivity studies to assess the applicability of the baseline TWCF results to PWRs in general:

- a sensitivity study assessing the method used to simulate increased levels of embrittlement

- a sensitivity study assessing the applicability of the baseline results on plate welded vessels to predict the TWCF of forged vessels

- the effect of vessel thickness on TWCF

The following sections summarize the motivation for and results of these sensitivity studies.

5.1 Simulating Increased Levels of Embrittlement

Use of more realistic models and input values than those used in the calculations that provide the technical basis for the current PTS rule produces a considerable reduction in the estimated values of TWCF. As detailed in 20, at 60 EFPYs (an operational lifetime beyond that anticipated after a single license extension), the TWCF values estimates for the three study plants lie between 10^{-11} and 10^{-8} events/year. However, the through-wall cracking frequency limit consistent with RG 1.174 is 10^{-6} events/year. Consequently, to develop reference-temperature-based screening limits, it was necessary to increase the level of embrittlement of the vessels for the three study plants so that the estimated TWCF values would approach the 10^{-6} events/year limit. In the baseline calculations reported in 20, embrittlement was artificially increased by increasing EFPY (increasing time) and by extrapolating fluence in linear proportion to time. An alternative procedure for artificially increasing embrittlement would be to allow the temporal and irradiation exposure parameters to remain within realistic ranges and instead increase the unirradiated transition temperature (the $RT_{NDT(u)}$) of the beltline materials. To determine what effect these two procedures have on the estimated TWCF values, the NRC performed sensitivity studies using the Beaver Valley and Palisades plants. In these sensitivity studies, the 32 EFPY analyses reported in 20 were treated as a baseline above which embrittlement was increased. Relative to this baseline, each EFPY/time increase can be expressed as an increase in the reference temperature by subtracting the reference temperature associated with 32 EFPY from the reference temperature associated with a particular EFPY/time increment. In this sensitivity study, the NRC compared the TWCF increases produced by these EFPY/time-driven reference temperature increases with TWCF increases driven by simply increasing the $RT_{NDT(u)}$ of the beltline materials by different fixed amounts. Figure 5-1 shows the result of this analysis, which demonstrates that the EFPY/time method of artificially increasing embrittlement results in TWCF estimates that exceed those produced by the alternative method of increasing $RT_{NDT(u)}$.

It must be emphasized that both of these procedures (as well as any other alternative procedure) extrapolate outside of the empirical bounds of the database used to establish the embrittlement shift model. The staff selected the EFPY/time extrapolation method over the $RT_{NDT(u)}$ extrapolation method for the baseline calculations reported in 20 because the embrittlement shift model includes explicitly both time and irradiation exposure variables. During the development of this model, the staff considered the known physical bases for time/exposure trends, and incorporated this knowledge into the functional form of the model [17]. Thus, there is some reason to expect that time and irradiation exposure variables will extrapolate better than the fracture toughness before irradiation begins (as quantified by $RT_{NDT(u)}$), which was not considered in the development of the embrittlement shift model.

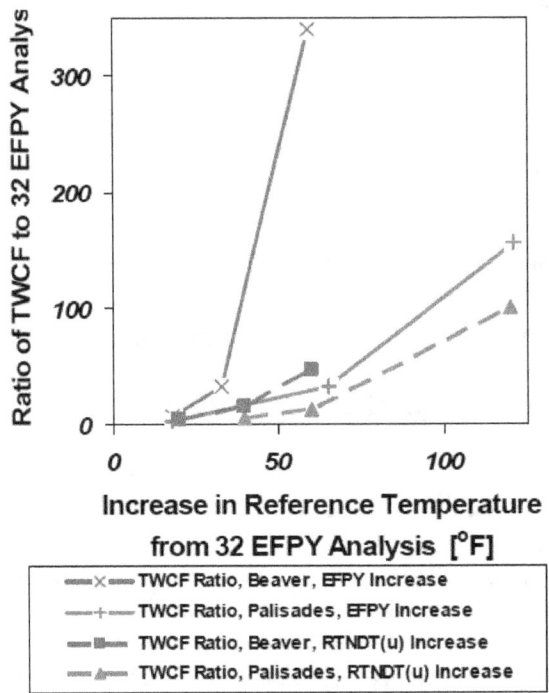

Figure 5-1 Effect of different methods to artificially increase embrittlement on the predicted TWCF values

5.2 Applicability to Forged Vessels

All three of the study vessels are fabricated by welding together plates that are roll-formed to make either 120° or 180° segments. However, 21 of the operating PWRs have beltline regions made of ring forgings. As such, these vessels have no axial welds. The lack of the large axially oriented axial weld flaws in such vessels indicates that they should, in general, have much lower values of TWCF than a comparable plate vessel of equivalent embrittlement. However, forgings have a population of embedded flaws that is particular in density and size to their method of manufacture. Additionally, under certain conditions, forgings are subject to subclad cracking associated with the deposition of the austenitic stainless steel cladding layer. Thus, to investigate the applicability of the baseline results reported in 20 to forged vessels, the NRC performed a number of analyses on vessels using properties (e.g., $RT_{NDT(u)}$, Cu, Ni, P) and flaw populations appropriate to forgings.

Appendix B details the technical basis for the distributions of flaws used in these sensitivity studies. The distribution of embedded forging flaws is based on destructive examination of an RPV forging [37]. These flaws are similar in both size and in density to plate flaws. The distribution of subclad cracks is based on a review of the literature on subclad flaws, particularly a summary article by Dhooge [8]. Subclad cracks occur as dense arrays of shallow cracks extending into the vessel wall from the clad-to-base metal interface to depths limited by the heat-affected zone (approximately 2 mm). These cracks are oriented normally to the direction of welding for clad deposition, producing axially oriented cracks in the vessel beltline. They are clustered where the passes of strip clad contact each other. Subclad flaws are much more likely to occur in particular grades of pressure vessel steels that have chemical compositions that enhance the likelihood of cracking. Forging grades such as A508 are more susceptible than plate materials such as A533. High levels of heat input during the cladding process also enhance the likelihood of subclad cracking.

5.2.1 Embedded Forging Flaw Sensitivity Study

The NRC used the following three steps to construct this sensitivity study:

(1)   Two sets of forging properties were selected—those of the Sequoyah 1 and Watts Bar 1 RPVs [31]. These properties were selected because they are among the most irradiation sensitive of all the forging materials in RVID.

(2)   Two hypothetical models of forged vessels were constructed based on the existing models of the Beaver Valley and Palisades vessels. In each case, the hypothetical forged vessels were constructed by removing the axial welds and combining these regions with the surrounding plates to make "forgings." These forgings were assigned the material properties from Step 1.

(3)   A FAVOR analysis of each vessel/forging combination from Steps 1 and 2 was conducted at two embrittlement levels, 32 EFPY and Ext-B. Thus, a total of $2^3 = 8$ FAVOR analyses were performed (2 material property definitions x 2 vessel definitions x 2 embrittlement levels).

On average, the TWCF of the forging vessels was only 3 percent of the plate welded vessels, and at most it was 15 percent. These reductions are consistent with those expected when the large axial weld flaws are removed from the analysis.

5.2.2 Subclad Crack Sensitivity Study

The NRC used the following three steps to construct this sensitivity study:

(1)   One set of forging properties was selected—that of the Sequoyah 1 [31].

(2)   One hypothetical model of a forged vessel was constructed based on the existing model of the Beaver Valley. The hypothetical forged vessel was constructed by removing the axial welds and combining these regions with the surrounding plates to make a forging. This forging was assigned the properties from Step 1.

(3)   A FAVOR analysis of each vessel/forging combination from Steps 1 and 2 was conducted at three embrittlement levels, 32 EFPY, 60 EFPY, and Ext-B. Thus, a total of 3 FAVOR analyses was performed (1 material property definition x 1 vessel definition x 3 embrittlement levels).

At 32 and 60 EFPY, the TWCF of the forging vessels was approximately 0.2 percent and 18 percent of the plate welded vessels. However, at the much higher embrittlement level represented by the Ext-B condition, the forging vessels had TWCF values 10 times higher than that characteristic of plate welded vessels at an equivalent level of embrittlement. While these very high embrittlement levels are unlikely to be approached in the foreseeable future, these results indicate that a more detailed assessment of vessel failure probabilities associated with subclad cracks would be warranted should a subclad-cracking-prone forging be subjected to very high embrittlement levels in the future.

Table 5-1 Results of Forging Flaw Sensitivity Study

Baseline Plate Welded Plant	Forging Property Set	EFPYs	Base FCI	Sensitivity FCI	FCI ratio	Base TWCF	Sensitivity TWCF	TWCF ratio
Palisades	1	32	9.67e-8	4.24e-9	0.04	9.98e-9	3.68e-12	0.004
	2	32	9.67e-8	2.08e-8	0.22	9.98e-9	2.31e-12	0.000
	1	Ext-b	5.80e-6	3.67e-6	0.06	2.10e-6	7.02e-8	0.03
	2	Ext-b	5.80e-6	6.76e-7	0.12	2.10e-6	9.93e-9	0.005
Beaver Valley	1	32	1.56e-7	1.76e-8	0.11	1.40e-9	1.94e-11	0.014
	2	32	1.56e-7	5.67e-9	0.04	1.40e-9	3.91e-12	0.003
	1	Ext-Bb	9.00e-6	2.44e-6	0.27	3.81e-7	5.86e-8	0.15
	2	Ext-Bb	9.00e-6	4.59e-7	0.05	3.81e-7	7.98e-9	0.021

Forging Property Set 1: Cu = 0.13%; Ni = 0.76%; P = 0.020%; $RT_{NDT(u)}$ = 73 F;USE = 72 ft-lbs (Sequoyah)
Forging Property Set 2: Cu = 0.17%' Ni = 0.80%; P = 0.012%; $RT_{NDT(u)}$ = 47 F;USE = 50 ft-lbs (Watts Bar)

Table 5-2 Results of Subclad Crack Sensitivity Study

EFPYs	Base FCI	Forging subclad flaws FCI	FCI ratio subclad /base	Base TWCF	Forging subclad flaws TWCF	TWCF ratio subclad /base
32	1.56e-7	1.60e-8	0.10	1.40e-9	2.57e-12	0.0018
60	5.66e-7	9.60e-8	0.17	6.15e-9	1.09e-9	0.18
Ext-Bb	9.00e-6	1.31e-5	1.46	3.81e-7	3.95e-6	10.37

The baseline for all analyses was Beaver Valley as reported by [EricksonKirk 04b]. The sensitivity analyses are based on forging property set 1, as defined in Table 5-1

5.3 Effect of RPV Wall Thickness on TWCF

20 noted in the FAVOR results for primary-side pipe breaks a potential effect of vessel wall thickness on the conditional probability of through-wall cracking. This effect can be expected for the following reasons:

- The magnitude of thermal stress scales in proportion to the thickness, with thicker vessels generating higher levels of thermal stress. Figure 5-2 shows the effect of this increased thermal stress on the applied driving force to fracture associated with a large-diameter pipe break. This effect will tend to increase the probability of through-wall cracking for thicker vessels.

- Because thicker vessels will have a larger volume of plate material and a larger weld fusion line area they will also have a larger number of flaws. This effect will also tend to increase the probability of through-wall cracking for thicker vessels.

- There is more distance in a thicker vessel over which an initiated crack can arrest, thereby not failing the vessel. Also, thicker vessels would tend to have more weld layers with different copper contents. This effect will tend to reduce the probability of through-wall cracking for thicker vessels.

To investigate the effect of these first two factors (the third could not be investigated without modifying the structure of the FAVOR code), the NRC increased the thickness of the Beaver Valley vessel from 7 7/8 in. (its actual thickness) in five increments up to 11 in. (characteristic of the thickest PWRs in service; see Figure 5-3). For each of these five thicker versions of Beaver Valley, the NRC used FAVOR to estimate the CPTWC of the following four transients (all of which are dominant contributors to the TWCF of Beaver Valley):

- BV9—16-in. diameter hot-leg break

- BV56—4-in. diameter surge line break

- BV126—stuck-open safety relief valve that recloses after 100 minutes resulting in repressurization of the primary system

- BV102—main steamline break

Figure 5-4 shows that increasing the vessel wall thickness increases the CPTWC for all four transients. Recalling that these CPTWC values would be weighted by their initiating event frequencies (and those of other transients) to obtain a TWCF estimate, these results suggest that, through a wall thickness of 9.5 in. (thicker than all but three of the in-service PWRs), the integrated effect of wall thickness on TWCF should be modest (i.e., a factor of approximately 3 increase at most) relative to the analyses of the three study plants (one 7 7/8-in. thick vessel and two 8.5-in. thick vessels). For vessels of greater wall thicknesses, a plant-specific analysis is warranted to properly capture all aspects of increased vessel wall thickness on TWCF. However, given that the three plants of 11 in. and greater thickness are Palo Verde Units 1, 2, and 3, and that these vessels have very low embrittlement projected at either EOL or end of license extension, the practical need for such plant-specific analysis is mitigated. It can also be noted that use of the baseline TWCF results from the three study plants will overestimate the TWCF of the seven thinner operating PWRs (7 in. thick or less).

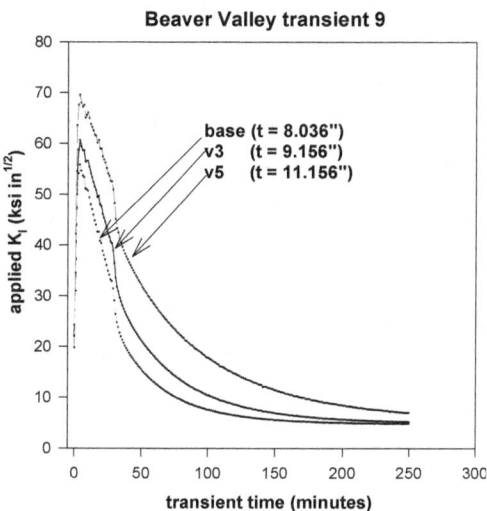

Figure 5-2 Effect of vessel wall thickness on the variation of applied-K_I vs. time for a 16-in. diameter hot-leg break in Beaver Valley. The flaw has the following dimensions: L=0.35 in., 2a=0.50 in, 2c=1.5 in.

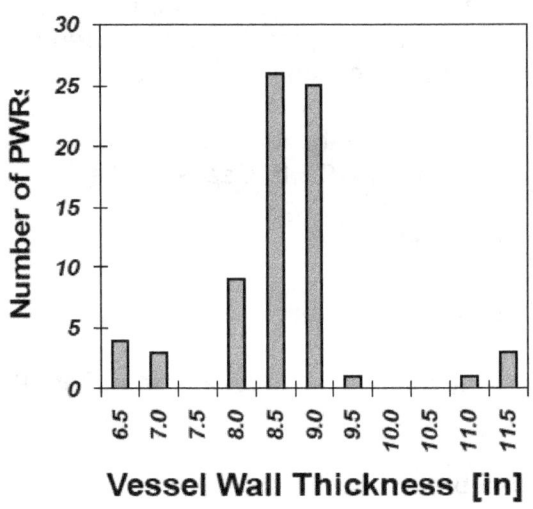

Figure 5-3 Distribution of RPV wall thicknesses for PWRs currently in service [31]

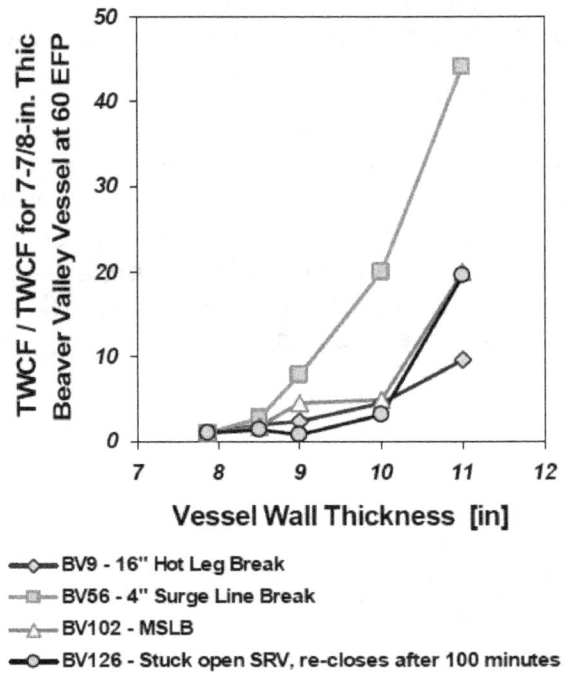

=◇= BV9 - 16" Hot Leg Break
—☐— BV56 - 4" Surge Line Break
—△— BV102 - MSLB
=○= BV126 - Stuck open SRV, re-closes after 100 minutes

Figure 5-4 Effect of vessel wall thickness on the TWCF of various transients in Beaver Valley (all analyses at 60 EFPYs)

6. SUMMARY AND RECOMMENDATIONS

This report documents sensitivity studies performed on the FAVOR PFM model (and on PFM-related variables) with two goals in mind:

- To provide confidence in the robustness of the PFM model, the NRC has assessed the effect of the following credible model and input perturbations on TWCF estimates:

  — residual stresses assumed to exist in the RPV wall

  — embrittlement shift model

  — resampling of chemical composition variables at the 1/4 T, 1/2 T, and 3/4 T locations for welds

  — upper-shelf toughness model

- To provide confidence that the results of the calculations for three specific plants can be generalized to apply to all PWRs, the NRC performed the following sensitivity studies to assess the influence of factors not fully considered in the baseline TWCF estimates:

  — method for simulating increased levels of embrittlement
  — assessment of the applicability of these results to forged vessels
  — effect of vessel thickness

In the former category, all effects were negligible or small. The small effects included the adoption of an embrittlement shift model that differs from that in 3 (which increases TWCF by approximately a factor of 3). In addition, the NRC model accounts for distinctly different copper contents in different weld layers (which reduces TWCF by approximately a factor of 2.5 relative to the assumption that the mean value of copper does not vary through the vessel thickness). Neither of these effects is significant enough to warrant a change to the baseline model or to recommend a caution regarding its robustness.

Sensitivity studies in the latter category suggest the following minor cautions regarding the applicability of the baseline results for the three study plants reported in 20 to PWRs in general:

- In general, the TWCF of forged PWRs can be assessed using the baseline results (see NUREG-1806 [20]) by ignoring the TWCF contribution of axial welds. However, should changes in future operating conditions result in a forged vessel being subjected to very high levels of embrittlement, a plant-specific analysis to assess the effect of subclad flaws on TWCF would be warranted.

- For PWR vessels ranging in thickness from 7.5 in. to 9.5 in., the baseline TWCF results are realistic. The baseline results overestimate the TWCF of the seven thinner vessels (i.e., wall thicknesses below 7 in.) and underestimate the TWCF of Palo Verde Units 1, 2, and 3, all of which have wall thicknesses above 11 in. However, these thicker vessels have very low embrittlement projected at either EOL or end of license extension, suggesting little practical effect of this underestimation.

7. REFERENCES

1. *U.S. Code of Federal Regulations*, Title 10, Section 50.61, "Fracture Toughness Requirements for Protection Against Pressurized Thermal Shock Events."

2. American Society of Mechanical Engineers, *1998 ASME Boiler and Pressure Vessel Code*, "Rules for Construction of Nuclear Power Plants, Division 1," Subsection NB, Class 1 Components, ASME NB-2331.

3. American Society for Testing and Materials, "Standard Guide for Predicting Radiation-Induced Transition Temperature Shift in Reactor Vessel Materials, E706 (IIF)," ASTM E900-02, ASTM, 2002.

4. Bryan, R. H., et al., "Pressurized Thermal Shock Test of 6-in.-Thick Pressure Vessels. PTSE-1: Investigation of Warm-Prestressing and Upper-Shelf Arrest," NUREG/CR-4106 (ORNL-6135), Oak Ridge National Laboratory, April 1985.

5. J. W. Bryson and T. L. Dickson, "Stress-Intensity-Factor Influence Coefficients for Axial and Circumferential Flaws in Reactor Pressure Vessels," PVP Vol. 250, ASME Pressure Vessels and Piping Conference, pp. 77–88, 1993.

6. Chang, Y.H., et al., "Thermal Hydraulic Uncertainty Analysis in Pressurized Thermal Shock Risk Assessment," University of Maryland, ADAMS Accession No. ML043570541.

7. Cheverton, R. D., et al., "Pressure Vessel Fracture Studies Pertaining to the PWR Thermal-Shock Issue: Experiments TSE-5, TSE-5A, and TSE-6," NUREG/CR-4249 (ORNL-6163), Oak Ridge National Laboratory, June 1985.

8. Dhooge, A., et al., "A Review of Work Related to Reheat Cracking in Nuclear Reactor Pressure Vessel Steels," *International Journal of Pressure Vessels and Piping*, Vol. 6, pp. 329–409, 1978.

9. Dickson, T. L., "Validation of FAVOR Code Linear Elastic Fracture Solutions for Finite-Length Flaw Geometries," PVP Vol. 304, *Fatigue and Fracture Mechanics in Pressure Vessels and Piping*, ASME, 1995.

10. Dickson, T.L., et al., "Evaluation of Margins in the ASME Rules for Defining the P-T Curve for a RPV," *Proceedings of the ASME Pressure Vessel and Piping Conference*, 1999.

11. Dickson, T. L., B.R. Bass, and P.T. Williams, "Validation of a Linear-Elastic Fracture Methodology for Postulated Flaws Embedded in the Wall of a Nuclear Reactor Pressure Vessel," PVP Vol. 403, *Severe Accidents and Other Topics in RPV Design*, American Society of Mechanical Engineering Pressure Vessels and Piping Conference, pp. 145–151, 2000.

12.   Dickson, T. L., B.R. Bass, and P.T. Williams, "A Comparison of Fracture Mechanics Methodologies for Postulated Flaws Embedded in the Wall of a Nuclear Reactor Pressure Vessel, Pressure Vessel and Piping Design Analysis," ASME Publication PVP Vol. 430, pp. 277–284, July 2001.

13.   Dickson, T.L., and F.A. Simonen, "The Impact of an Improved Flaw Model on a Pressurized Thermal Shock Evaluation," *Proceedings of the 2002 ASME Pressure Vessel and Piping Conference, August 2002*, Vancouver, Canada.

14.   Dickson, T.L., et al., "Preliminary Results of the United States Nuclear Regulatory Commission's Pressurized Thermal Shock Rule Reevaluation Project, " *Probabilistic Aspects of Life Prediction, ASTM STP 1450*, W.S. Johnson and B.M. Hillberry, eds., ASTM International, West Conshohocken, PA, 2003.

15.   Dickson, T.L. and S. Yin, "Electronic Archival of the Results of Pressurized Thermal Shock Analyses for Beaver Valley, Oconee, and Palisades Reactor Pressure Vessels Generated with the 04.1 version of FAVOR," ORNL/NRC/LTR-04/18, Adams Accession No. ML042960465.

16.   Eason, E.D., J.E. Wright, and E.E. Nelson, "Multivariable Modeling of Pressure Vessel and Piping *J-R* Data," U.S. Nuclear Regulatory Commission, NUREG/CR-5729, April 1991.

17.   Kirk, M.T., et al., "Updated Embrittlement Trend Curve for Reactor Pressure Vessel Steels," *Transactions of the 17th International Conference on Structural Mechanics in Reactor Technology (SMiRT 17)*, Prague, Czech Republic, August 17–22, 2003.

18.   English, C., and W. Server, "Attenuation in US RPV Steels—MRP-56," Electric Power Research Institute, June 2002.

19.   EricksonKirk, M., et al., "Probabilistic Fracture Mechanics: Models, Parameters, and Uncertainty Treatment used in FAVOR Version 04.1," NUREG-1807.

20.   EricksonKirk, M., et al., "Technical Basis for Revision of the Pressurized Thermal Shock (PTS) Screening Limit in the PTS Rule (10 CFR 50.61): Summary Report," NUREG-1806.

21.   EricksonKirk, M.A., "Materials Reliability Program: Implementation Strategy for Master Curve Reference Temperature, T_o (MRP-101)," Electric Power Research Institute and the U.S. Department of Energy, Washington, DC, 1009543, 2004.

22.   Kirk, M., and M.E. Natishan, "Shift in Toughness Transition Temperature Due to Irradiation: ΔT_o vs. ΔT_{41J}, A Comparison and Rationalization of Differences," *Proceedings of the IAEA Specialists Meeting on Master Curve Technology,* Prague, Czech Republic, September 2001.

23. Kirk, M., "Inclusion of Warm Pre-Stress Effects in Probabilistic Fracture Mechanics Calculations Performed to Assess The Risk of RPV Failure Produced by Pressurized Thermal Shock Events: An Opinion," NATO Science Series Volume xxx, *Scientific Fundamentals for the Lifetime Extension of Reactor Pressure Vessels*, Hans-Werner Viehrig, ed., 2002.

24. Kirk, M.T., et al., "Technical Basis for Revision of the Pressurized Thermal Shock (PTS) Screening Criteria in the PTS Rule (10 CFR 50.61)," U.S. Nuclear Regulatory Commission, ADAMS Accession No. ML030090626, December 2002.

25. "An Assessment of the Integrity of PWR Vessels," 2nd report by a study group under the chairmanship of D.W. Marshall, published by the UKAEA, 1982.

26. Meyer, T., et al., "Materials Reliability Program: Pressurized Thermal Shock Sensitivity Studies Using the FAVOR Code (MRP-96)," Electric Power Research Institute, November 2003.

27. Memorandum from Thadani to Collins on "Transmittal of Technical Work to Support Possible Rulemaking on a Risk-Informed Alternative to 10 CFR 50.46/GDC 35," ADAMS Accession No. ML022120660, July 31, 2002.

28. Oak Ridge National Laboratory, "Pressurized Thermal Shock Evaluation of the Calvert Cliffs Unit 1 Nuclear Power Plant," NUREG/CR-4022 (ORNL/TM-9408), September 1985.

29. Oak Ridge National Laboratory, "Pressurized Thermal Shock Evaluation of the H.B. Robinson Unit 2 Nuclear Power Plant," NUREG/CR-4183 (ORNL/TM-9567), September 1985.

30. Oak Ridge National Laboratory, "Preliminary Development of an Integrated Approach to the Evaluation of Pressurized Thermal Shock as Applied to the Oconee Unit 1 Nuclear Power Plant," NUREG/CR-3770 (ORNL/TM-9176), May 1986.

31. Nuclear Regulatory Commission Reactor Vessel Integrity Database, Version 2.1.1, July 6, 2000.

32. U.S. Nuclear Regulatory Commission, "Format and Content of Plant-Specific Pressurized Thermal Shock Safety Analysis Reports for Pressurized Water Reactors," Regulatory Guide 1.154, 1987.

33. U.S. Nuclear Regulatory Commission, "Thermal Annealing of Reactor Pressure Vessel Steels," Regulatory Guide 1.162.

34. U.S. Nuclear Regulatory Commission, "Calculational and Dosimetry Methods for Determining Pressure Vessel Neutron Fluence," Regulatory Guide 1.190, March 2001.

35.   U.S. Nuclear Regulatory Commission, "Radiation Embrittlement of Reactor Vessel Materials," Regulatory Guide 1.99, Version 2, February 1986, ADAMS Accession No. ML003740284.

36.   Sakamoto, K., et al., "Development of Prediction Equations on Charpy Upper Shelf Energy for Japanese RPV Steels," *Proceedings of the 2003 ASME Pressure Vessel and Piping Symposium, Cleveland, Ohio,* August 2003.

37.   Schuster, G.J., "Technical Letter Report —JCN-Y6604—Validated Flaw Density and Distribution Within Reactor Pressure Vessel Base Metal Forged Rings," prepared by Pacific Northwest National Laboratory for the U.S. Nuclear Regulatory Commission, December 20, 2002.

38.   U.S. Nuclear Regulatory Commission, "Pressurized Thermal Shock," SECY-82-465, November 23, 1982.

39.   U.S. Nuclear Regulatory Commission, "Status Report: Risk Metrics and Criteria for Pressurized Thermal Shock," SECY-02-0092, May 30, 2002.

40.   Simonen, F.A., et al., "A Generalized Procedure for Generating Flaw Related Inputs for the FAVOR Code," NUREG/CR-6817, Rev. 1.

41.   "Uncertainty Analysis and Pressurized Thermal Shock, An Opinion," U.S. Nuclear Regulatory Commission, 1999, ADAMS Accession No. ML992710066.

42.   Sokolov, M.A., and R.K. Nanstad, "Comparison of Irradiation Induced Shifts of K_{Jc} and Charpy Impact Toughness for Reactor Pressure Vessel Steels," *ASTM STP-1325,* American Society of Testing and Materials, 1996.

43.   Sokolov, M.A., and R.K. Nanstad, "Comparison of Irradiation Induced Shifts of K_{Jc} and Charpy Impact Toughness for Reactor Pressure Vessel Steels," NUREG/CR-6609, November 2000.

44.   Wallin, K., T. Saario, and K. Törrönen, "Statistical Model for Carbide Induced Brittle Fracture in Steel," *Metal Science,* Vol. 18, pp. 13–16, January 1984.

45.   Wallin, K., et al., "Mechanism-Based Statistical Evaluation of the ASME Reference Fracture Toughness Curve," *5th International Conference on Pressure Vessel Technology,* Vol. II, "Materials and Manufacturing," ASME, San Francisco, CA, 1984.

46.   Wallin, K., "The Scatter in K_{Ic} Results," *Engineering Fracture Mechanics,* Vol. 19, No. 6, pp. 1085–1093, 1984.

47.   Wallin, K., "The Size Effect in K_{Ic} Results," *Engineering Fracture Mechanics,* Vol. 22, pp. 149–163, 1985.

48.   Wallin, K., "Statistical Modeling of Fracture in the Ductile to Brittle Transition Region," Defect Assessment in Components— Fundamentals and Applications," ESIS/EGF9, J.G. Blauel and K.H. Schwalbe, eds., pp. 415–445, 1991.

49.   Wallin, K., "Irradiation Damage Effects on the Fracture Toughness Transition Curve Shape for Reactor Vessel Steels," *Int. J. Pres. Ves. & Piping*, Vol. 55, pp. 61–79, 1993.

50.   Wallin, K., "Statistical Aspects of Constraint with Emphasis on Testing and Analysis of Laboratory Specimens in the Transition Region," *Constraint Effects in Fracture, ASTM STP-1171,* E.M. Hackett, K.H. Schwalbe, and R.H. Dodds, eds., American Society for Testing and Materials, 1993.

51.   Wallin, K., "Loading Rate Effect on the Master Curve T_o," Paper IIW- X-1403-97, 1997.

52.   Wallin, K., "Master Curve Analysis of Ductile to Brittle Region Fracture Toughness Round Robin Data: The 'EURO' Fracture Toughness Curve," VTT Manufacturing Technology, VTT Publication 367, 1998.

53.   Wallin, K., and R. Rintamaa, "Master Curve Based Correlation Between Static Initiation Toughness K_{Ic} and Crack Arrest Toughness K_{Ia}," *Proceedings of the 24th MPA-Seminar, October 8 and 9, 1998,* Stuttgart.

54.   Wallin, K., "Low-cost *J-R* curve estimation based on CVN upper shelf energy," *Fatigue and Fracture of Engineering Materials and Structures*, Vol. 24, pp. 537–549, 2001.

55.   Whitehead, D.W., et al., "Generalization of Plant-Specific Pressurized Thermal Shock (PTS) Risk Results to Additional Plants," ADAMS Accession No. ML042880476.

56.   Williams, P.T. and T.L. Dickson, "Fracture Analysis of Vessels—Oak Ridge, FAVOR v04.1, Computer Code: Theory and Implementation of Algorithms, Methods, and Correlations," NUREG/CR-6854.

57.   Woods, R., et al., "Selection of Pressurized Thermal Shock Transients to Include in PTS Risk Analysis," *IJPVP*, Vol. 78, pp. 179–183, 2001.

APPENDIX A

FLAW DISTRIBUTION CORRESPONDENCE

The report authored by F.A. Simonen, S.R. Doctor, G.J. Schuster, and P.G. Heasler entitled, "A Generalized Procedure for Generating Flaw Related Inputs for the FAVOR Code," also known as NUREG/CR-6817, Revision 1, published in October 2003, details the flaw distribution adopted in FAVOR and used in this investigation (Simonen 10-03).

This appendix includes the text of a letter sent to the primary author of this report, Dr. Fredric Simonen, and Dr. Simonen's reply. The letter clarifies Dr. Simonen's views regarding the extent to which the flaw distributions reported in NUREG/CR-6817, Revision 1, apply to operating PWRs in general.

Text of Letter Sent to Dr. Simonen

30th June 2004

MEMORANDUM

From: Mark EricksonKirk (mtk@nrc.gov)
To: Fred Simonen (fredric.simonen@pnl.gov)

cc: Debbie Jackson
       Allen Hiser

Subj: NUREG/CR-6817, Rev. 1, "A Generalized Procedure for Generating Flaw-Related Inputs for the FAVOR Code," by F. A. Simonen, et al.

Motivated by comments received from both the external peer review panel we convened for the PTS project and from some members of the industry I have recently re-read the subject NUREG/CR report. For the PTS re-evaluation effort it is important to know to what extent the flaw distributions reported in NUREG/CR-6817, Rev. 1 apply to operating PWRs *in general*. Neither the executive summary nor the conclusions of this report (which I have attached for your reference) speak to this issue. However, I did find the following statements in the body of the report that speak to the question of the general applicability of the flaw distribution:

On p. 5.9 (*emphasis* added):

> The PRODIGAL model provided a systematic approach to relate flaw occurrence rates and size distributions to the parameters of welding processes that can vary from vessel-to-vessel. Application of the model showed the sensitivity of calculated flaw distributions to changes in the welding process conditions. ***Calculations with PRODIGAL and consideration of known differences in fabrication procedures used to manufacture U.S. vessels indicated that data from PVRUF and Shoreham can reasonably be applied to all vessels at U.S. plants.***

On p. 6-2 (*emphasis* added)

Use of Data Versus Models and Expert Elicitation - In developing flaw distributions, measured data were used to the maximum extent possible. The PRODIGAL flaw simulation model and results of the expert judgment elicitation were used only when the data were inadequate. In the case of seam welds, there was a relatively large amount of data, and the PRODIGAL model and expert elicitation were not used to quantify estimates of flaw densities and sizes. The PRODIGAL model did, however, suggest the normalization of flaw dimension by the dimensions of weld beads and the separation of data into subsets corresponding to small and large flaws (as defined by flaw depth dimensions relative to the weld bead dimensions). *In addition, the expert elicitation and the PRODIGAL model helped to justify the application of data from the PVRUF and Shoreham vessels to the larger population of vessels at U.S. nuclear plants.*

The NUREG/CR also includes the following statement:

On p. 6-3 (*emphasis* added)

Vessel-to-Vessel Variability - The PNNL examinations of vessel material focused on two vessels (PVRUF and Shoreham), with only limited examinations of material from other vessels (Hope Creek, River Bend, and Midland). The Shoreham flaws showed some clear differences from the PVRUF flaws with somewhat greater flaw densities and longer flaws (larger aspect ratios). However, there was no basis for relating these differences in flaw densities and sizes to other vessels. With only two examined vessels it was not possible to statistically characterize vessel-to-vessel differences such that the differences could be simulated as a random factor in Monte Carlo calculations. The decision was to develop separate procedures to generate flaw distributions for the PVRUF and Shoreham vessels. *Following the conservative approach taken in other aspects of the PTS evaluations where data and/or knowledge is lacking, it was recommended that the Shoreham version of the flaw distribution be used in PTS calculations, which served to ensure conservatism in the predictions of vessel failure probabilities.*

The statements from p. 5-9 and 6-2 suggest that the view that the flaw distributions proposed in NUREG/CR-6817, Rev. 1 apply to operating PWRs *in general*. Conversely, the statement made on p. 6-3 seems to suggest that you and your co-authors view the flaw distributions as being *conservative*.

To help me respond to questions I have received regarding use of the flaw distributions presented in the NUREG/CR in the PTS re-evaluation project it would be most helpful to me if you could respond to the following question:

What is the view of you and your co-authors? Do you view the flaw distributions published in NUREG/CR-6817, Rev. 1 as being applicable to PWRs in general, or do you view them as being a conservative representation of the flaw population in the fleet of operating PWRs.

I greatly appreciate your assistance with this matter.

Reply Received from Dr. Simonen

>>> "Simonen, Fredric A" <fredric.simonen@pnl.gov> 07/01/04 02:23PM >>>

Mark:

This is my response to the questions that you posed to me in the attached memo (June 30, 2004):

What is the view of you and your co-authors? Do you view the flaw distributions published in NUREG/CR-6817, Rev. 1 as being applicable to PWRs in general, or do you view them as being a conservative representation of the flaw population in the fleet of operating PWRs?

Your June 30, 2004 memo accurately reflects my views and those of my co-authors regarding the applicability of the flaw distributions in NUREG/CR-6817, Rev1 to PWRs in general as well as the conservative nature of the distributions.

In developing the flaw distribution methodology we were guided by Lee Abramson (statistician from NRC staff) in dealing with uncertainties. Because the PNNL flaw data were primarily from two vessels (PVRUF and Shoreham) a rigorous statistical treatment of vessel-to-vessel differences was not possible. The flaw model was therefore developed with separate treatments for the two vessels, along with a recommendation to use the more conservative treatment based on the Shoreham vessel when addressing other vessels. Other conservatisms can be introduced as appropriate in application of the flaw model to address uncertainties in knowledge regarding of a specific vessel. One example of such uncertainties would be the amount of repair welding in a particular vessel.

Fredric A. Simonen
Laboratory Fellow
Pacific Northwest National Laboratory
P.O. Box 999
2400 Stevens Drive
Richland, Washington 99352 USA
phone 509-375-2087
fax   509-375-6497
email fredric.simonen@pnl.gov

  <<d jackson memo 30 jun 04.doc>>

APPENDIX B

FLAW DISTRIBUTIONS FOR FORGINGS

This appendix includes two articles prepared by Dr. Frederic Simonen of the Pacific Northwest National Laboratory concerning flaw distributions in forgings. The staff used these articles as the basis of the forging flaw sensitivity studies reported in Section 5.2.

Technical Basis for the Input Files to FAVOR Code for Flaws in Vessel Forgings

F.A. Simonen
Pacific Northwest National Laboratory
Richland, Washington

July 28, 2004

Pacific Northwest National Laboratory (PNNL) has been funded by the U.S. Nuclear Regulatory Commission (NRC) to generate data on fabrication flaws that exist in reactor pressure vessels (RPV). Work has focused on flaws in welds but with some attention also to flaws in the base metal regions. Data from vessel examinations along with insights from an expert judgment elicitation (MEB-00-01) and from applications of the PRODIGAL flaw simulation model (NUREG/CR-5505, Chapman et. al. 1998) have been used to generate input files (see report NUREG/CR-6817, Simonen et. al. 2003) for probabilistic fracture mechanics calculations performed with the FAVOR code by Oak Ridge National Laboratory. NUREG/CR-6817 addresses only flaws in plate materials and provided no guidance for estimating the numbers and sizes of flaws in forging materials. More recent studies have examined forging material, which has provided a data on flaws that were detected and sized in the examined material. At the request of NRC staff PNNL has used these more recent data to supplement insights from the expert judgment elicitation to generate FAVOR code input files for forging flaws. The discussion below describes the technical basis and results for the forging flaw model.

Nature of Base Metal Flaws

PNNL examined material from some forging material from a Midland vessel as described by Schuster (2002). The forging was made during 1969 by Ladish. Examined material included only part of the forging that had been removed from the top of the forged ring as scrap not intended for the vessel. This material was expected to have more than the average flaw density, and as such may contribute to the conservatism of any derived flaw distribution.

Figures 1 and 2 show micrographs of small flaws in plate and forging materials. These flaws are inclusions rather than porosity or voids. They are also not are planer cracks. Therefore their categorization as simple planar or as volumetric flaws is subject to judgment. The plate flaw of Figure 1 has many sharp and crack-like features, whereas such features are not readily identified for the particular forging flaw seen in Figure 2. It should, however, be emphasized that the PNNL examined only a limited volume of both plate and forging material and found very few flaws in examined material. It is not possible to generalize from such a small sample of flaws. Accordingly, the flaw model makes assumptions that may be somewhat conservative, due to the limited data on the flaw characteristics.

Flaw Model for Forging Flaws

The model for generating distributions of forging flaws for the FAVOR code uses the same approach as that for modeling plate flaws as described in NUREG/CR-6817. The quantitative results of the expert elicitation are used along with available data from observed forging flaws. The flaw data were used as a "sanity check" on the results of the expert elicitation. Figure 3 summarizes results of the expert elicitation. Each expert was asked to estimate ratios between flaw densities in base metal compared to the corresponding flaw densities observed in the weld metal of the PVRUF vessel. Separate ratios were requested for plate material and forging material.

As indicated in Figure 3, the parameters for forging flaws are similar to those for plate flaws. The forging and plate models used the same factor of 0.1 for the density of "small" flaws (flaws with through-wall dimensions less than the weld bead size of the PVRUF vessel). The density of "large" flaws in forging material is somewhat greater than the density of flaws in plate material. The factor of 0.025 for the flaw density is replaced by a factor of 0.07 for forging flaws. A truncation level of 0.11 mm is used for both plate and forging flaws. As described in the next section the data from forging examinations show that these factors are consistent with the available data. It is noted that the assumption for the 0.07 factor is supported by only a single data point corresponding to the largest observed forging flaw (with a depth dimension of 4 mm).

The factors of 0.1 and 0.07 came from the recommendations from the expert elicitation on vessel flaws. As noted below the very limited data from PNNL's examinations of forging material show that these factors are consistent with the data, although the 0.07 factor is supported by only one data point for an observed forging flaw with a 4-mm depth dimension.

Comparison with Data on Observed Flaws

The PNNL examinations of vessel materials included both plate materials and forging materials. For plate flaws less than 4-mm in through-wall depth dimension, Figure 4 shows data from NUREG/CR-6817 that shows frequencies for plate flaws. Also shown for comparison are the flaw frequencies for the welds of the PVRUF and Shoreham vessels. This plot confirmed results of the expert judgment elicitation (Figure 4) and indicated: 1) there are fewer flaws in plate material than in weld material, and 2) there is about a 10:1 difference in flaw frequencies for plates versus welds.

PNNL generated the data on flaws in forgings after preparation of NUREG/CR-6817. Forging data are presented in Figures 5 and 6 along with the previous data for flaws in the PVRUF plate material. There is qualitative agreement with the results of the expert judgment elicitation (Figure 4), which indicates that 1) plate and forging materials have similar frequencies for small (2 mm) flaws, and 2) forging material have higher flaw frequencies for larger (>4 mm) flaws.

Inputs for FAVOR Code

Figure 7 compares the flaw frequencies for plates and forgings that were provided to ORNL as input files for the FAVOR code. This plot shows mean frequencies from an uncertainty distribution as described by the flaw input files. It is seen that the curves for plate and forging flaws are identical for small flaws but show differences for the flaws larger than 3 percent of the vessel wall thickness. Also seen is the effect of truncating the flaw distribution at a depth of 11 mm (about 5 percent of the wall thickness).

References

Jackson, D.A. and L. Abramson, 2000. *Report on the Preliminary Results of the Expert Judgment Process for the Development of a Methodology for a Generalized Flaw Size and Density Distribution for Domestic Reactor Pressure Vessel*, MED-00-01, PRAB-00-01,U.S. Nuclear Regulatory Commission.

Schuster, G.J., 2002. "Technical Letter Report – JCN-Y6604 – Validated Flaw Density and Distribution Within Reactor Pressure Vessel Base Metal Forged Rings," prepared by Pacific Northwest National Laboratory for U.S. Nuclear Regulatory Commission, December 20, 2002.

Simonen, F.A., S.R. Doctor, G.J. Schuster and P.G. Heasler, 2003. *A Generalized Procedure for Generating Flaw-Related Inputs for the FAVOR Code*, NUREG/CR-6817, Rev. 1, prepared by Pacific Northwest National Laboratory for U.S. Nuclear Regulatory Commission.

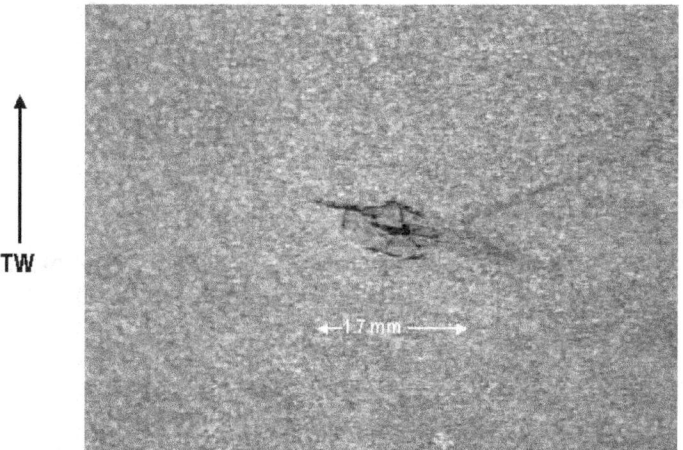

Figure 1 Small Flaw in Plate Material

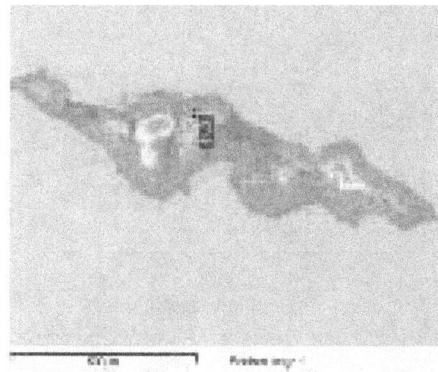

Figure 2 Small Flaw in Forging Material

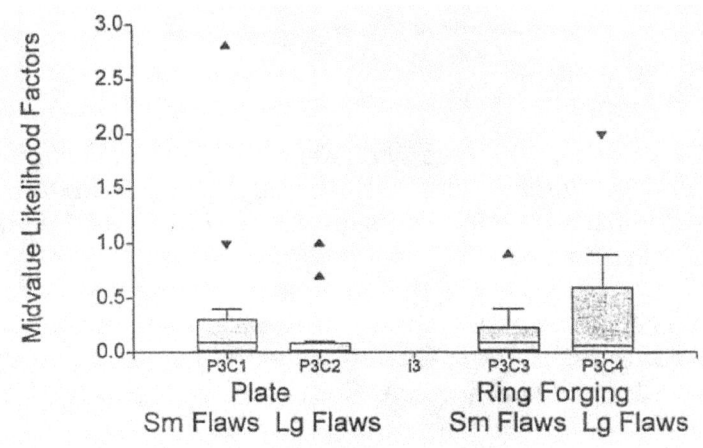

	Base Metal vs. Weldmetal			
	Plate vs. Welds		Ring Forgings vs. Welds	
	Small Flaws	Large Flaws	Small Flaws	Large Flaws
MIN	.0004	.001	.001	.002
LQ	.015	.01	.02	.007
MED	.1	.025	.1	.07
UQ	.3	.09	.2	.6
MAX	12.0	1.0	.9	2.0

Figure 3   Relative Flaw Densities of Base Metal Compared to Weld Metal as Estimated by Expert Judgment Process (from Jackson and Abramson 2000)

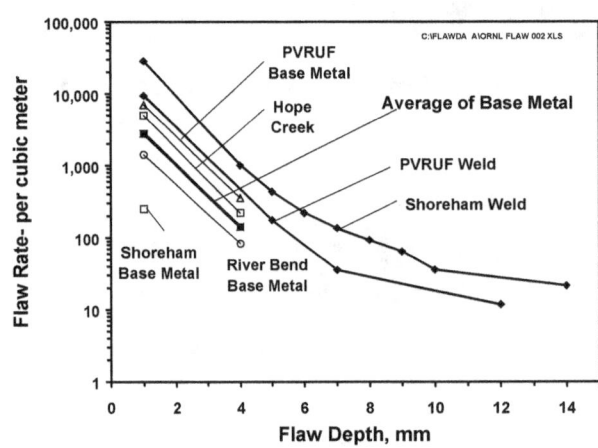

Figure 4 Flaw Frequencies for Plate Materials with Comparisons to Data for Weld Flaws

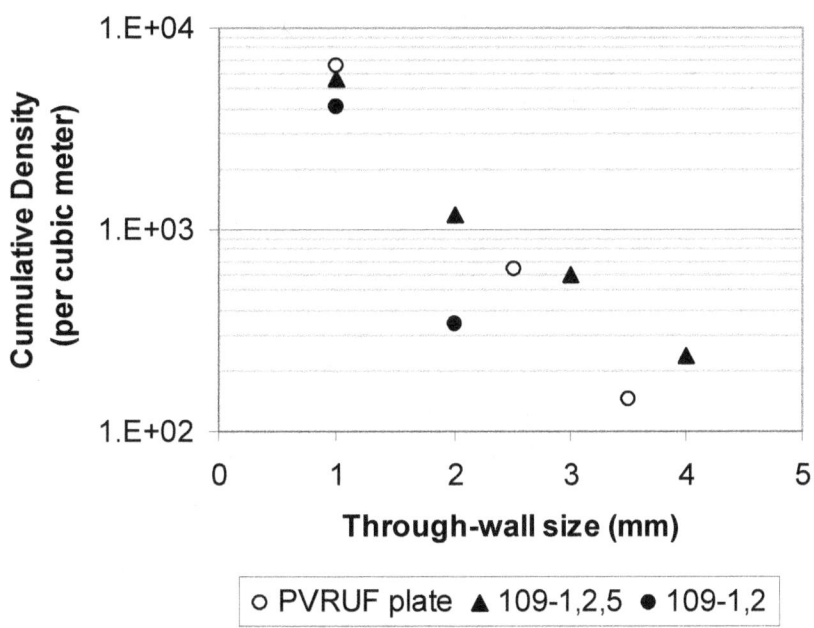

Figure 5 Validated Flaw Density and Size Distribution for Three Forging Specimens. Cumulative flaw density is the number of flaws per cubic meter of equal or greater size.

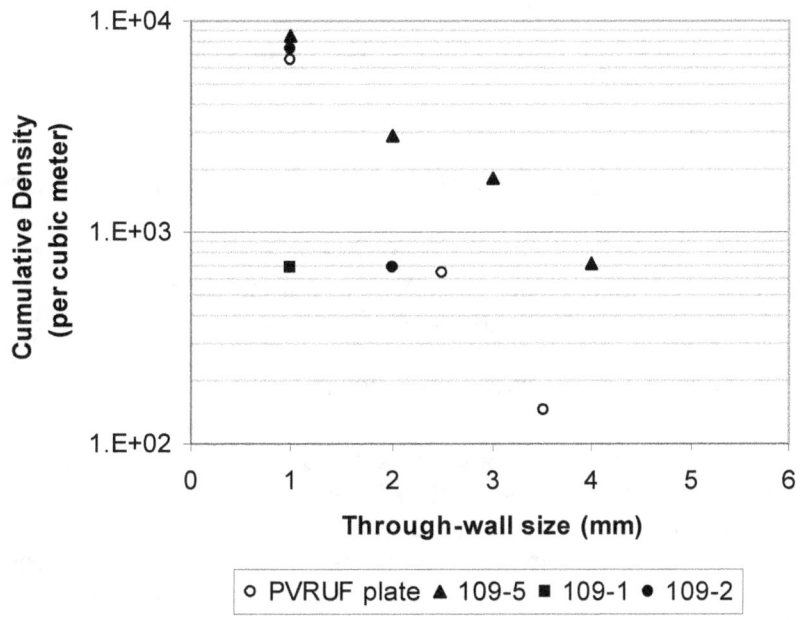

Figure 6 Average of Validated Cumulative Flaw Density for Forging Material, A508

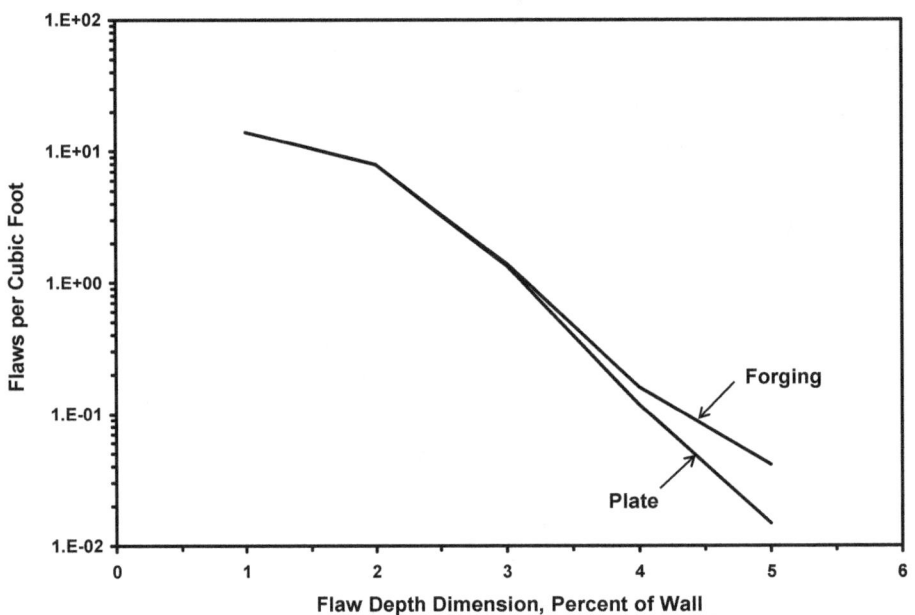

Figure 7 Comparison of Flaw Distributions for Forging and Plate

---

Basis for Assigning Subclad Flaw Distributions

F.A. Simonen
Pacific Northwest National Laboratory
Richland, Washington

September 29, 2004

Pacific Northwest National Laboratory (PNNL) has supported the U.S. Nuclear Regulatory Commission (USNRC) in the efforts to revise the Pressurized Thermal Shock (PTS) Rule. In this role PNNL has provided Oak Ridge National Laboratory (ORNL) with inputs to describe the distributions of fabrication flaws in reactor pressure vessels. These inputs, consisting of computer files, have been a key input to the probabilistic fracture mechanics code FAVOR. Flaw inputs have addressed seam welds, cladding and base metal materials, but had specifically excluded subclad flaws associated with the heat affected zone (HAZ) from the process that deposits stainless steel cladding to the inner surface of the vessel. Recently ORNL was requested by USNRS to evaluate the potential contribution of these subclad flaws to reactor pressure vessel failure for PTS conditions. The present paper describes the technical basis for the subclad flaw input files that PNNL provided to ORNL for use with the FAVOR code.

PNNL has examined material from vessels welds, basemetal and cladding and has used the data on observed flaws in these material types to establish statistical distributions for the numbers and sizes of flaws in these categories of materials. None of the examined material showed any evidence of the type of subclad flaws of interest. Therefore, the numbers and sizes of sub clad flaws for a vessel susceptible to such cracking was estimating from a review of the literature. The primary source was a comprehensive paper summarizing European work from the 1970's (A. Dhooge, R.E. Dolby, J. Sebille, R. Steinmetz ad A.G. Vinckier, "A Review of Work Related to Reheat Cracking in Nuclear Reactor Pressure Vessel Steels", International Journal of Pressure Vessels and Piping, Vol. 6, 1978, pp.329-409). This paper was based on experience with vessel cracking in Europe and subsequent research programs conducted during the 1970's. The paper should therefore be relevant to US concerns with older vessels that may have been fabricated with European practices.

The literature shows that subclad cracks 1) are shallow flaws extending into the vessel wall from the clad-to-basemetal interface with 4-mm being cited as a bounding through-wall depth dimension, 2) have orientations normal to the direction of welding for clad deposition - giving axial cracks in a vessel beltline, 3) occur as dense arrays of small cracks extending into the vessel wall from the clad to basemetal interface, 4) extend to depths limited by the heat affected zone. Pictures in the cited paper show networks of cracks with typical depths estimated from micrograph being significantly less than the bounding 4-mm depth. The cracks occur perpendicular to the direction of welding and are clustered where the passes of strip clad contact each other. Subclad flaws are much more likely to occur in particular grades of pressure vessel steels that have chemical compositions that enhance the likelihood of cracking. Forging grades such as A508 Class 2 are more susceptible than both A507 Class 3 forgings and plate materials such as A533B. High levels of heat input during the cladding process also enhance the likelihood of subclad cracking. In addition other details of the cladding process are important such as single layer versus two layer cladding.

The numbers of cracks per unit area of vessel inner surface were estimated from Figure 1 of the Dhooge paper. Cracking was shown to occur in bands estimated to have a width of 4 mm. This dimension was used to estimate a bounding length of subclad cracks. The longest individual cracks in Figure 1 were about 2-mm versus the 4-mm width dimension of the zone of cracking. By counting the number of cracks pictured in small region of vessel surface crack density of 80,512 flaws per square meter was estimated.

The flaw input files as provide to ORNL were based on the following assumptions:

1. The crack depth dimensions were described by a uniform statistical distribution from 0 to 4 mm with no cracks greater than 4 mm in depth.

2. The crack lengths were also described by a uniform statistical distribution. Like our assumption for flaws in seam welds, the amount by which flaw lengths exceed their corresponding depth dimension is taken to be a uniform distribution from 0 to 4 mm. Thus the extreme length for a flaw with a depth of 4 mm is 8 mm. The 4 mm deep flaws can therefore have lengths ranging from 4 to 8 mm (aspect ratios from 1:1 to 2:1). Flaws with depths of 1 mm can have lengths ranging from 1 mm to 5 mm (aspect ratios from 1:1 to 5:1).

3. The flaw density expressed as flaws per unit area was converted (for purposes of the FAVOR code) to flaws per unit volume based on the total volume of the metal in the vessel wall.

4. The file prepared for FAVOR assumes that the code simulates flaws for the total vessel wall thickness, rather than just the category 1 and 2 regions which address only the inner 3/8 of the wall thickness. Terry will need to account for this concern during the FAVOR calculations

The resulting very large number of flaws (> 130,000) per vessel is based on a photograph of one small area of a vessel surface, with the implication that this area was representative of the entire vessel. It is possible that subclad flaws tend to occur in patches of the vessel surface. However it is generally understood that subclad flaws occur in a wide spread manner and that there are very large numbers of flaws given the conditions for subclad cracking exist. Based on PNNL's limited review of documents it is difficult to justify reducing the estimated flaw density of subclad flaws. However, it would be useful to perform a sensitivity calculation to see if refinement of my estimate would have a significant effect on the FAVOR calculations.

The estimated depths of the subclad flaws is probably conservative. The depth of 4 mm was based on statements regarding bounding flaw depths of 4 mm, with no other evidence such as micrographs of flaws or data on measured depth dimensions presented. The depth of 4 mm could be an estimate for the size of the heat affected zone, which is then taken as a limitation on flaw depth. Alternatively the 4 mm depth could be the extreme measured (or reported depth) of some observed subclad flaws. A review of available papers showed some examples from metallography of subclad flaws, which showed only flaws of much smaller depths (< 2 mm). It is therefore suggested that a sensitivity study be based on assumed subclad flaws with a bounding depth of 2 mm. In this case the FAVOR code would include only flaws in the "first bin" corresponding to sizes 0 to 1% of the vessel wall thickness.

PNNL's estimates of subclad flaw distributions have been based on a rather limited review of available literature with a particular focus on the Dhooge 1978 paper. It is recommended that the scope of the literature review be expanded and that individuals (domestic and overseas) be contacted to seek sources of additional information. PNNL will also review notes from past sessions with expert elicitation panels that have addressed reactor vessel fabrication and flaw distributions for the USNRC. The critical issue is the depth dimensions of typical subclad flaws. The depth dimension of 4 mm may be uncharacteristic of most subclad flaws, but rather a bounding dimension based on consideration of the depth of heat affected zones. This depth has also been used in the literature for deterministic fracture mechanics calculations and could therefore reflect the conservative nature of inputs used for such calculations.

NRC FORM 335 (9-2004) NRCMD 3.7	U.S. NUCLEAR REGULATORY COMMISSION	1. REPORT NUMBER (Assigned by NRC, Add Vol., Supp., Rev., and Addendum Numbers, if any.)
	BIBLIOGRAPHIC DATA SHEET *(See instructions on the reverse)*	NUREG-1808

2. TITLE AND SUBTITLE

Sensitivity Studies of the Probabilistic Fracture Mechanics Model Used in FAVOR

3. DATE REPORT PUBLISHED	
MONTH	YEAR
March	2010

4. FIN OR GRANT NUMBER

5. AUTHOR(S)

M.T. EricksonKirk[1], T.L. Dickson[2], T. Mintz[3], F.A. Simonen[4]

6. TYPE OF REPORT

Technical

7. PERIOD COVERED *(Inclusive Dates)*

6-1999 to 6-2006

8. PERFORMING ORGANIZATION - NAME AND ADDRESS *(If NRC, provide Division, Office or Region, U.S. Nuclear Regulatory Commission, and mailing address; if contractor, provide name and mailing address.)*

[1]Division of Fuel, Engineering, and Radiological Research, Office of Nuclear Regulatory Research, U.S. Nuclear Regulatory Commission, Washington, DC 20555-0001, [2]Oak Ridge National Laboratory, P.O. Box 2008, Oak Ridge, TN 37831-6075, [3]Southwest Research Institute, 9503 W Commerce, San Antonio, Texas 78227-1301, [4]Pacific Northwest National Laboratory, P.O. Box 999, 2400 Stevens Drive, Richland, Washington 99352 USA

9. SPONSORING ORGANIZATION - NAME AND ADDRESS *(If NRC, type "Same as above"; if contractor, provide NRC Division, Office or Region, U.S. Nuclear Regulatory Commission, and mailing address.)*

Division of Fuel, Engineering, and Radiological Research, Office of Nuclear Regulatory Research, U.S. Nuclear Regulatory Commission, Washington, DC 20555-0001

10. SUPPLEMENTARY NOTES

11. ABSTRACT *(200 words or less)*

During plant operation, the walls of reactor pressure vessels (RPVs) are exposed to neutron radiation, resulting in localized embrittlement of the vessel steel and weld materials in the core area. If an embrittled RPV had a flaw of critical size and certain severe system transients were to occur, the flaw could very rapidly propagate through the vessel, resulting in a through-wall crack and challenging the integrity of the RPV. The severe transients of concern, known as pressurized thermal shock (PTS), are characterized by a rapid cooling (i.e., thermal shock) of the internal RPV surface in combination with repressurization of the RPV. Advancements in our understanding and knowledge of materials behavior, our ability to realistically model plant systems and operational characteristics, and our ability to better evaluate PTS transients to estimate loads on vessel walls led the U.S. Nuclear Regulatory Commission (NRC) to realize that the earlier analysis, conducted in the course of developing the PTS Rule in the 1980s, contained significant conservatisms.

This report, which describes sensitivity studies performed on the probabilistic fracture mechanics model, is one of a series of 21 other documents detailing the results of the NRC study.

12. KEY WORDS/DESCRIPTORS *(List words or phrases that will assist researchers in locating the report.)*

Pressurized thermal shock, reactor pressure vessel, probabilistic fracture mechanics

13. AVAILABILITY STATEMENT
unlimited
14. SECURITY CLASSIFICATION
(This Page)
unclassified
(This Report)
unclassified
15. NUMBER OF PAGES
16. PRICE

NRC FORM 335 (9-2004)

PRINTED ON RECYCLED PAPER